First Edition

Genuine Autographed Collectible

Do you want me to sign it in ink or in lipstick?

GENIUS: A GIFT OF DIVINE REVELATION

NEW SCIENTIFIC THEORY
LAWS OF INEXTRICABILITY

Gift Card

Date:

To:

From:

Message:

LAWS OF INEXTRICABILITY
THE GRAND UNIFIED THEORY OF PHYSICS

1. PEACELESSNESS
2. HOMELESSNESS
3. LIFELESSNESS
4. HUMANBEINGLESSNESS
5. MORALITYLESSNESS
6. CHOICELESSNESS
7. LOVELESSNESS
8. MINDLESSNESS
9. HAPPYLESSNESS
10. MERCYLESSNESS
11. CREATIONLESSNESS
12. TRUTHLESSNESS
13. GODLESSNESS
14. RELIGIONLESSNESS
15. PROBLEMLESSNESS

Now You Know Everything About Everything

NEW SCIENTIFIC THEORY

THE GRAND UNIFIED THEORY OF PHYSICS

Our World Is Organized By Laws of Inextricability

What Do Books Do?

BOOKS ARE POWERFUL!

Books Educate!
Books Enlighten!
Books Empower!
Books Emancipate!
Books Entertain!
Books Spring Eternal!
Books Drive Exploration!
Books Spark Evolution!
Books Ignite Revolution!

Sharon Esther Lampert

THE END
OF FEEL GOOD
PSYCHOBABBLE TO
NUMB THE PAIN AND
KEEP THEM SANE!

ONLY GOD
KNOWS
WHY
AND GOD
KNOWS BEST!

Philosopher Queen Sharon Esther Lampert

Self-Help, Philosophy, Science, Physics, Religion, Genius, Sharon Esther Lampert

NOW YOU KNOW EVERYTHING ABOUT EVERYTHING
OUR WORLD IS ORGANIZED BY LAWS OF INEXTRICABILITY
THE GRAND UNIFIED THEORY OF PHYSICS

KADIMAH PRESS
GIFTS OF GENIUS

Books may be purchased for education, business, or sales promotional use.

ISBN Hardcover: 979-8-3492-9552-2
ISBN Paperback: 979-8-3492-9555-3
ISBN E-Book: 979-8-3492-9553-9
Library of Congress Catalog Card Number: 2025908023

FAN MAIL:
www.SharonEstherLampert.com   FANS@SharonEstherLampert.com

Cover and Interior Book Design: Creative Genius Sharon Esther Lampert

Editor: Dave Segal

Palm Beach Book Publisher
Phone: 561-888-0313
Sharon@PalmBeachBookPublisher.com

To Order Book:
Ingram, 1 Ingram Blvd. La Vergne, TN 37086-3629
Phone: 615-793-5000
Fax orders: 615-287-6990

First Edition

Manufactured in the United States of America

Now You Know Everything About Everything

NEW SCIENTIFIC THEORY
THE GRAND UNIFIED THEORY OF PHYSICS

Our World Is Organized By Laws of Inextricability

BE ART
ART IS SMART
ART IS OF THE HEART
MAKE ART NOT WAR
YOU ARE BORN FOR GREATNESS
YOU ARE A MASTERPIECE

Sharon Esther Lampert

To My Gift of Divine Revelation

Thank You for the Gift of Genius; It Is the Seed in Me that I Was Born to Be

MOMMY

LOVE OF MY LIFETIME
WHO KNEW WHO I WAS
FROM THE **INSIDE OUT**!

Age 9
THE QUEEN HAS ARRIVED!
My daughter is a poet, a philosopher
and a teacher.
She is the Princess & the Pea!
BEAUTY & BRAINS!
LOVE & XOXO
MOMMY

BEZALEL: EXODUS 31: 1-3
31 Then the Lord said to Moses,
2 "See, I have chosen Bezalel son
of Uri, the son of Hur, of the tribe of
Judah, 3 and I have filled him with
the Spirit of God, with wisdom, with
understanding, with knowledge
and with all kinds of skills."

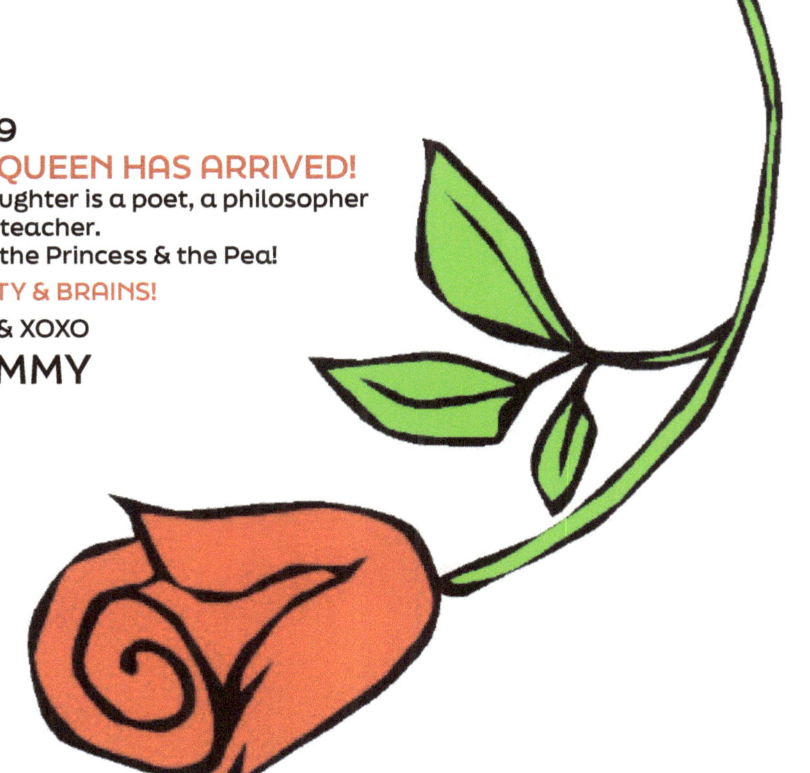

Dearest Sharon Esther:

I Know That I Know Nothing!

SOCRATES
PHILOSOPHER KING

Dearest Socrates:

ONE GLOBAL ENEMY: IGNORANCE

Sharon Esther Lampert
PHILOSOPHER QUEEN

THE SOCRATIC METHOD

Dearest Sharon Esther:

"An Unexamined Life Is Not Worth Living

My Way Toward Truth Is to Ask the Right Questions:

Q: Is Life Worth Living?"

SOCRATES
PHILOSOPHER KING

THE TALMUDIC METHOD

Read My Book:
THE WHY CHILD

Dearest Socrates:

I asked and answered 50+ questions in 80+ books.

HIGHLIGHTS:

Q1: Billions of people have live and died:
Why is the planet still a torture chamber of
unspeakable horrors?

Q2: **HOW SCARY IS THAT? THE DARK TRIAD**

Q3: What would the world look like if educators
knew how to nurture and cultivate the awesome
power of the human brain? (A: SMARTGRADES)

Q4: Is life a gift or a punishment? (A: **LAWS OF INEXTRICABILITY**)

Q5: Measure the pain? (life is a punishment)

Q6: Measure the pleasure? (life is a gift)

Q7: How do live a moral life in an immoral universe?

Q8: Why do bad things happen to good people? (A: **8 OUTCOMES**)

Q9: Why is it **NORMAL** for billions of people to talk to **GOD**—but **CRAZY** if
GOD talks to me? (A: GOD IS GO! DO!)

Q10: What is the greatest **LIE** ever told in the name of **GOD**?

Q11: What is love?

Q12: Why is the person you love and married the same person you
hate and divorced?

Q13: What happens when you dress up Albert Einstein as Marilyn Monroe?
(A: **SHARON ESTHER LAMPERT**)

Sharon Esther Lampert
PHILOSOPHER QUEEN

PARDES: 4 LEVELS
1. PSHAT (simple)
2. REMEZ (hints)
3. DRASH (context)
4. SOD (hidden)

Dearest Sharon Esther:

"To a Mind That Is Still The Whole Universe Surrenders."
— Lau Tzu

Dearest Lau Tzu:

LAWS OF INEXTRICABILITY
THE GRAND UNIFIED THEORY OF PHYSICS

Sharon Esther Lampert
PHILOSOPHER QUEEN

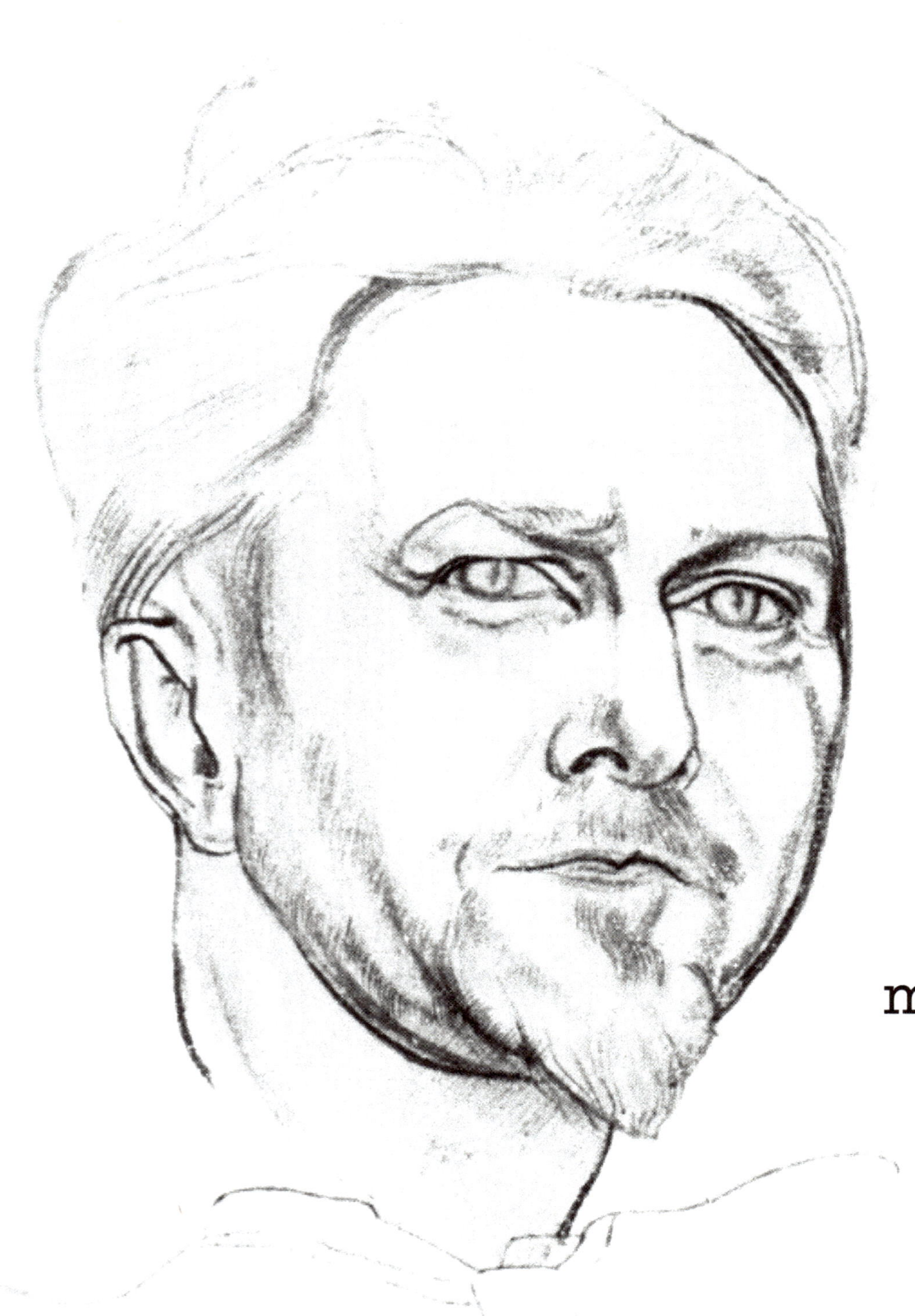

Dearest
Sharon Esther:

"**GENIUS**...
is the
capacity
to see
ten things
where the
ordinary
man sees one."

Poet Ezra Pound

Dearest Ezra:

THE WORLD IS ORGANIZED BY LAWS OF INEXTRICABILITY

1. PEACELESSNESS
2. HOMELESSNESS
3. LIFELESSNESS
4. HUMANBEINGLESSNESS
5. MORALITYLESSNESS
6. CHOICELESSNESS
7. LOVELESSNESS
8. MINDLESSNESS
9. HAPPYLESSNESS
10. MERCYLESSNESS
11. CREATIONLESSNESS
12. TRUTHLESSNESS
13. GODLESSNESS
14. RELIGIONLESSNESS
15. PROBLEMLESSNESS

Sharon Esther Lampert
PHILOSOPHER QUEEN

Dearest Sharon Esther:

"Talent Hits a Target
No One Else Can Hit
Genius Hits a Target
No One Else Can See."

— Arthur Schopenhauer

Dearest Arthur:

You can compete with the **GOOD** and the **GREAT**—but not with the **GIFTED!**

GENIUS: THE GIFT OF DIVINE REVELATION

SHARON ESTHER LAMPERT
PRODIGY: EXODUS 31:1-3

GENIUS: GIFTS OF DIVINE REVELATION

Read: THE SPERM MANIFESTO: THERE IS ONLY ROOM FOR ONE AT THE TOP

Read: Who Knew **GOD** Was Such a Chatterbox: A Working Definition of **GOD**

Read: Unleash The Creator The **GOD** Within: 10 Esoteric Laws of Genius and Creativity

Read: The Awesome Art of Alliteration Using One Letter of the Alphabet (6 Books)

Read: **SMARTGRADES BRAIN POWER REVOLUTION** (25+ Books)

Read: WOMEN HAVE ALL THE POWER BUT HAVE NEVER LEARNED HOW TO USE IT

Read: THE 22 COMMANDMENTS: A UNIVERSAL MORAL COMPASS

Dearest Sharon Esther:

"Imagination is more important than knowledge.

Knowledge is limited.

Imagination encircles the world.

Logic will get you from A to B. Imagination will take you everywhere.

Creativity is seeing what others see, and thinking what no one else has ever thought."

— Albert Einstein

Dearest Albert:

MY BOOKS WRITE THEMSELVES!

I wake-up in the middle of the night and write the whole book!

Read: UNLEASH THE CREATOR GOD WITHIN
10 ESOTERIC LAWS OF GENIUS AND CREATIVITY

1. V.E.S.S.E.L.
2. INSPIRATION
3. IMPREGNATION
4. INCUBATION
5. GENESIS
6. SILENT: LISTEN
7. METAMORPHOSIS
8. REVELATION
9. SIGNATURE
10. IMMORTALITY

SHARON ESTHER LAMPERT
PRODIGY: EXODUS 31:1-3

V.E.S.S.E.L. **VERY. EXTRA. SPECIAL. SHARON. ESTHER. LAMPERT.**

Who Knew God Was Such a Chatterbox?

Dearest Sharon Esther,
Do you catch my drift?

And God Said: Go to Work!

GOD IS GO! DO!
GOD CAN ONLY DO FOR YOU WHAT GOD CAN DO THROUGH YOU

God Is Not Physics—The Laws of the Universe

GOD IS METAPHYSICS!
An Invisible and Intangible Entity—Like Your Mind, Thoughts, and Ideas
Beyond The Scope of Scientific Inquiry!

PRAY As If It All Depends On **GOD!**
WORK As If It All Depends On **YOU!**

Dearest GOD:

"If I was a Muslim,
they would chop off my head in a heartbeat.
If I was a Christian,
they would banish me to spend an eternity
burning in the inferno of Hell.
But it is the privilege of a Jew to
be able to quarrel with God."

SHARON ESTHER LAMPERT
PHILOSOPHER QUEEN
ISRAEL: TO STRUGGLE WITH GOD
GENESIS: 32:28
PRINCESS KADIMAH: 8TH PROPHETESS OF ISRAEL

Dearest Sharon Esther:

"Life is not a problem
to be solved
but a reality to
be experienced."

Søren Kierkegaard

Dearest Søren Kierkegaard:

Now You Know Everything About Everything

NEW SCIENTIFIC THEORY
THE GRAND UNIFIED THEORY OF PHYSICS

Our World Is Organized By Laws of Inextricability

Dearest Sharon Esther:

"Don't bend; don't water it down; don't try to make it logical; don't edit your soul according to the fashion. Rather, follow your most intense obsessions mercilessly."

Franz Kafka

Table of Contents

A GEM AMONG GIANTS
Measure the Pain!
Life is a Punishment
Measure the Pleasure
Life is a Gift
15 Esoteric Laws of Inextricability
ONLY GOD KNOWS WHY? AND GOD KNOWS BEST!
THE END OF FEEL GOOD PSYCHOBABBLE TO NUMB THE PAIN AND KEEP THEM SANE!
Q: Is Life a Gift or a Punishment?
15 Absolute Truths

Preface

DISRAELI said, "Philosopy becomes poetry and science imagination, in the enthusiasm of genius." This book is a confluence of philosophic genius and poetic genius. It is philosophy condensed into philosophical poetry.

Every noteworthy topic in philosophy is thematically interwoven into a set of unified, seamless, universal, and absolute truths, entitled, **15 ESOTERIC LAWS OF INEXTRICABILIT**Y, and every single truth, is a timeless truth for all eons.

Finding one philosophical truth is outstanding, but locating, compiling, and explaining 15 philosophical truths is beyond belief.

It has taken generations for a **PHILOSOPHER QUEEN** to be born of this caliber, who can reduce our reality into single words, that can communicate complex philosophical concepts, that both students and scholars, will immediately find comprehensible, not to mention endlessly thought provoking: (1) **Peace**lessness; (2) **Home**lessness; (3) **Life**lessness; (4) **Human**beinglessness; (5) **Morality**lessness; (6) **Choice**lessness; (7) **Love**lessness; (8) **Mind**lessness; (9) **Happy**lessness; (10) **Mercy**lessness; (11) **Creation**lessness (12) **Truth**lessness; (13) **GOD**lessness; (14) **Religion**lessness; and (15) **Problem**lessness. Here it is: THE GRAND UNIFIED THEORY OF PHYSICS. The single thread intertwining these terms, the philosophical concepts, is the poetic genius at work.

CONFUCIUS said to his pupil, "Do you think I have come to know many things by studying them? No. I penetrate them by their underlying unity."

Depth Perception: The **PHILOSOPHER QUEEN** in our midst, has revealed to us a perception of reality that is not relative, and that few people have ever laid their eyes, minds, and hearts upon, and has compiled her brilliant vision into one compact book.

LEO TOLSTOY said, "There is no genius where there is not simplicity."

In concurrence, **ALBERT EINSTEIN** said, "When the solution is simple, God is answering." This book will do for philosophy what Einstein did for physics. The "**15 LAWS OF INEXTRICABILITY**" will advance our understanding of our world, and give us a new paradigm of critical knowledge that we can no longer continue to live without. Effulgent, we now see the world we live in from new dimensions: "**15 LAWS OF INEXTRICABILITY**" are culled from an intuitive gift, with no high-tech microscopic or telescopic tools at her disposal.

Throughout history, women have been praised for their deft innate instincts, commonly referred to as, "**women's intuition.**" Finally, we bear witness to the tremendous power unleashed by one woman's intuition that is embodied by this magnificent and monumental achievement.

ROBERT GRAVES said, "Intuition is the supra-logic that cuts out all the routine processes of thought and leaps straight from the problem to the answer." This philosophy book is the official vade mecum to attain an astute awareness of reality: A GEM AMONG GIANTS

GOD OF WHAT?
Measure the Pain!
Life is a Punishment
15 Esoteric Laws of Inextricability
Measure the Pleasure
Life is a Gift
ONLY GOD KNOWS WHY? AND GOD KNOWS BEST!
THE END OF FEEL GOOD PSYCHOBABBLE TO NUMB THE PAIN AND KEEP THEM SANE!
Is Life a Gift or a Punishment?
15 Absolute Truths

Introduction

15 ESOTERIC LAWS OF INEXTRICABILITY

For sure, 8-billion people do need the protective services of a supernatural God:

"Dear God, Help Me, Heal Me, and Protect Me from Harm!"

We were conceived without our consent or consideration. We are forced into being by an act of sexual intercourse.

Upon exiting the cozy womb, and inhaling our first breath, we are immediately scooped up, bathed, weighed, and given a swig of breast milk.

We are completely at the mercy of our parents—not a **God**.

In childhood, we are introduced to a supernatural God who created the heavens and the earth.

We are taught to worship this **God**, as a means of extracting from this **God** the wherewithal of maintaining our sanity, security, and survival.

In every part of the world, 8-billion people pray to a God or Gods, e.g., Muslims pray 5X a day; Jews pray 3X a day; Hindus perform puja 2X-3X a day.

Praying to a GOD for SALVATION is a time-consuming endeavor. Every hour on the hour, our SURVIVAL is in jeopardy: WHY? THE 15 ESOTERIC LAWS OF INEX-TRICABILITY.

In lieu of a connection with a God or Gods, secular people prefer meditation as the means to foster a connection with oneself for inner well being to achieve focus, clarity, resolve, and resilience before engaging with the demands of the outer world:

1. We are born hungry and must find a living being or plant to kill and eat.

2. We are born thirsty and must purify the contaminated water.

3. We are born naked and must sew our clothes.

4. We are born in a womb and must build a shelter.

5. We are born ignorant, unconscious, and irrational, and must seek knowledge, develop consciousness, and apply reason.

Q: How can one ever come to terms with one's life?

Q: Is life a gift or a punishment?

Q: Is one's journey worth the effort required to maintain one's existence?

Q: Wherein lies the secret of the mystery of our existence?

The Bible? The Tanakh? The Talmud? The Zohar? The Dead Sea Scrolls? The Koran? The Gospels? The Apocrypha? The Pseudepigrapha? The Mahabharata? The Book of The Dead? The Granth? The Avesta? The Dhammapada? The Five Classics? The Vedas? The Tao-te-Ching? The Tripitaka?

Q: How interesting is the actual experience of a life?

Here comes the easy part of life: Is waking up each day to go to the bathroom, put makeup up, eat breakfast, go to the bathroom, go to work, eat lunch, go to the bathroom, nosh on a fat-free snack, go back to work, eat dinner, go to the bathroom, remove my makeup, sneak a nosh on a high-caloric snack, get in bed, make love, go to sleep—day in and day out—encircling a twenty-four hour clock, seven days week, 365 days a year, a cause to celebrate or a prison sentence?

On any given day, I am quite dizzy from putting on my lipstick, reapplying my lipstick, and removing my lipstick!

Here comes the hard part of life: The experience of a life is a constant state of war between life and death, between knowledge and ignorance, between health and illness, between good and evil, and between love and hatred. The fight between these opposing dualities is a unrelenting daily struggle. To maintain our survival, our razor sharp swords must always be drawn, and never sleep in their sheaths:

- Farmers fight the noble fight against famine
- Doctors fight the noble fight against illness
- Teachers fight the noble fight against ignorance
- Politicians fight the noble fight against war
- Theologians fight the noble fight against hatred

This is why we are all so busy! Each and every day, and each and every hour of the day is a struggle to survive. Every arena of human endeavor is a monumental uphill climb that requires our unwavering attention. We are all so busy replenishing our daily sustenance: our work translates into money to pay for food, shelter, and clothing. There are a great many more hidden costs involved.

Biological sustenance is a daily struggle—and we are all engaged in fighting the noble fight for life itself.

Service to society: Life is interdependent labor-intensive enslavement.

And when we die, we finally have a chance to catch up on our lack of sleep from our exhausting lives—lives that we have had to expend exclusively to maintaining our biological survival: **LIFELESSNESS.**

In the here and now, we are designed not to live forever, but to maintain a temporary status, ill fated to weathering the dispassionate storms of life; storms without mercy. Our artificially contrived sense of unfairness, injustice, and inequality take their toll: **MERCYLESSNESS.**

Anger looms and lingers right below the surface of most people, from real stresses and imagined slights that we vent upon each other: **MINDLESSNESS.**

And betwixt and between, illness, disease, and premature death follows suit. Looking forward to our senior citizen years of half-price tickets at movies, museums, theatres, and on buses and subways—that is if you can find someone to go with you who is still alive and well enough to accompany you—is becoming a rarity, as heart attacks and cancer each take their turn devastating our loved ones: **CREATIONLESSNESS.**

Then there are the genetic predispositions and contagious diseases that threaten to take hold of us and cause senseless suffering, before those mythological golden years of debilitation and senility set in and ravage us:

" I can't walk well!"

" I can't see well!"

" I can't hear well!"

" I can't remember!"

Oh, the heart-wrenching drama of it all!

Inevitably, we have to die and of course, we have to die from something. So be it. Lucky is if you die in your sleep, after having had to take a nap after a tough game of golf, then, you are considered one of the most fortunate ones. Dying in the midst of doing something you love; dying without having to suffer any more than having had to lose that game of golf is a gift. And if you just have happened to have won that game of golf—what a send off!

Born here, or more matter of factly, forced here without our consent, and operating without a instruction manual, one must join in the daily struggle to maintain one's life, one's family, one's home, one's possessions—only to eventually forfeit it all.

Every material manifestation that we possess lives happily ever after in the home of an heir—or in the local thrift shop.

The futile fight for life ends in naught; they don't even bury you in your own underwear, unless, of course, you plan ahead and write a will, a sign of a truly enlightened being.

Therefore, if given a choice, would anyone want to be born at all?

WHY not let sleeping unborn souls that were resting in peace remain wherever they were and not wake them?

WHAT is it about the experience of a life that is so intriguing that we continue to march on and beat our drums, singing songs of adulation, while weathering the trials and tribulations of our arduous existence?

WHY do we fight the great fight to live here, a place littered with shattering minefields of myriad dimensions of evil incarnate: the daily blood bath of indigenous violence, both intrinsic and extrinsic, between plants, animals and humankind, indigenous natural disasters, and indigenous harrowing diseases?

Every single person is living on a private or public battlefield, and has to wrestle with insane **DEMONS** of one kind or another: insane workplace, insane illness, insane society, insane environment, insane spouse, insane child, insane parents, and insane self: MIND**LESSNESS.**

An authentic life, where dreams do come true, for most, not for just a minuscule fortunate few; where one's life story is imbued with peace, joy, love, brotherhood, justice, equality, freedom and harmony does not seem possible within these treacherous frontiers.

Few People Have Everything
Many People Have Something
Some People Have Nothing

Day in and day out, every hour of the day, if you make a wrong move... you will be doomed to death.

For instance, earthquakes, hurricanes, and wildfires threaten to annihilate us: HOMELESSNESS.

For instance, 24/7 invisible murdering mutating monster microbes threaten to annihilate us. They also infect us and turn us into killers. Vicious viruses infect us and force us to kill our loved ones—not our enemies.

WHY not have this planet that we call "HOME," despite the fact, that it will never ever feel like "HOME SWEET HOME" for all generations of all eons condemned as too dangerous and uninhabitable?

WELCOME TO PLANET EARTH
WARNING: LAWS OF INEXTRICABILITY
UNINHABITABLE! CONDEMED!

These are some of the questions to ponder as you read this book. The answers are nowhere to be found, however, after partaking of the philosophical insights unearthed in this philosophical gem, you may mature into a more insightful being than you were before you opened these pages of deft depths. I promise that you will drink from a full cup of enlightenment.

INEXTRICABILITY: It is meaningful to write about MEANINGLESSNESS

GENIUS: The Gift of Divine Revelation—My Books Write Themselves!

Read: WHO KNEW **GOD** WAS SUCH A CHATTERBOX! GOD IS GO! DO!

SHARON ESTHER LAMPERT
PRODIGY-PHYSICIST
September 22nd, 1999

15 ESOTERIC LAWS OF INEXTRICABILITY

1. **PEACELESSNESS:** INEXTRICABILITY OF LIFE & DEATH
   24/7 Bloodbath: You are Predator and Prey!

2. **LIFELESSNESS:** INEXTRICABILITY OF SURVIVAL & EXTINCTION
   24/7 Fight for Survival (Inextricability of Exertion & Recovery)

3. **HOMELESSNESS:** INEXTRICABILITY OF HOME & HOMELESS
   24/7 Planet Earth Is a Boobytrapped Torture Chamber of Unspeakable Horrors
   CURVEBALLS: EXPECT THE UNEXPECTED

4. **HUMANBEINGLESSNESS:** INEXTRICABILITY OF HUMAN ANIMAL & HUMANBEING
   Dark Ages: Talk Like a Humanbeing & Behave Like a Human Animal
   Enlightenment: Talk Like a Humanbeing & Behave Like a Humanbeing
   Humanbeing is an Artificial Construction of: Love, Learning, and Laws

5. **MORALITYLESSNESS:** INEXTRICABILITY OF GOOD & EVIL
   GOOD & EVIL Have Eight Unpredictable Outcomes

6. **CHOICELESSNESS:** INEXTRICABILITY OF DEFAULT & DESIGN
   Most of Your Life Is Out of Your Control — 100% of People Live Their Lives on a **DEFAULT** Setting
   Everyone Gets a Different Set of Cards in Life—What You Do With Your Cards Determines Your **DESTINY**

7. **LOVELESSNESS:** INEXTRICABILITY OF LOVE & HATE
   100% of People **HELP** You with Their Strengths & **HURT** You with Their Weaknesses
   100% of Relationships are **LOVE-HATE** Relationships

8. **MINDLESSNESS:** INEXTRICABILITY OF **THE DARK TRIAD**
   (1) Unconscious & Conscious; (2) Ignorant & Knowledgeable; and (3) Irrational & Rational
   THE DARK TRIAD: We Are Born Ignorant, Unconscious, and Irrational —We Live This Way! & We Die This Way!

9. **HAPPYLESSNESS:** INEXTRICABILITY OF PAIN & PLEASURE
   Example: Eating Cake Gives You Pleasure & Getting Fat from Eating Cake Gives You Pain

10. **MERCYLESSNESS:** INEXTRICABILITY OF DISADVANTAGE & ADVANTAGE (Global Feeding Frenzy)
    Your Disadvantage Is Another Person's Advantage.

11. **CREATIONLESSNESS:** INEXTRICABILITY OF CREATION & DESTRUCTION
    Creation Undergoes Decay, Deterioration, and Death
    For Instance, Ripe Lemons Provide Vitamin C—Rotten Lemons Make You Sick
    For Instance, Freshly Killed Carcass/Plant Sustains Your Life—Rotten Carcass/Plant Kills You by Food Poisoning

12. **TRUTHLESSNESS:** INEXTRICABILITY OF TRUTH & LIES = **HALF-TRUTHS**

13. **GODLESSNESS:** INEXTRICABILITY OF PHYSICAL & METAPHYSICAL WORLDS
    GOD IS GO! DO! GOD CAN ONLY DO FOR YOU WHAT GOD CAN DO THROUGH YOU

14. **RELIGIONLESSNESS:** INEXTRICABILITY OF MAKE-BELIEVERS & NON-BELIEVERS
    For Centuries: Make-Believers Terrorize Heretics, Apostates, and Infidels: Convert or Die! or Convert or Burn in Hell!

15. **PROBLEMLESSNESS:** INEXTRICABILITY OF PROBLEMS & SOLUTIONS
    Every Day You Arise to Solve Problems and Implement Solutions—Problem 1: "I'm Hungry"

LIVE TO FIGHT
FIGHT TO LIVE
BORN TO DIE
—SEL

THE GRAND UNIFIED THEORY OF PHYSICS

Socratic Method and Talmudic Method

Philosophical Questions of Inquiry:

Q1a: God of What? Q1b: What Is God the God Of?   A: GOD IS GO! DO!

Q2: Is Life a Gift or a Punishment? Measure the Pain? Measure the Pleasure?

Q3: Is one's journey worth the effort required to maintain one's existence?

Q4: Wherein lies the secret of the mystery of our existence?

Q5: How interesting is the actual experience of a life?

Q6: Is the 24/7 clock a prison sentence?

Q7: If given a choice, would anyone want to be born at all?

Q8: WHY not have this place that we call "HOME" condemned as dangerous and uninhabitable? (**LAW OF INEXTRICABILITY: HOMELESSNESS**)

Q9: Do our sufferings reflect the natural relationship that exists between all animals, of which we are a full-fledged member in good standing? (**LAW OF INEXTRICABILITY: HUMANBEINGLESSNESS**)

Q10: Is the other not more than mere lunch; the means of acquiring resources that maintain our survival? (**LAW OF INEXTRICABILITY: PEACELESSNESS**)

Q11: Do we invent our sufferings and perpetuate them by our ever fermenting irrational hatreds that we then vent upon our own specie? (**LAW OF INEXTRICABILITY: MINDLESSNESS**)

THE WHY CHILD:
Q12: WHY I was I born without my consent? Oy Vey!
Q13: WHY do I know these people: my parents? Oy Vey!
Q14: WHY did my cat have to die? Oy Vey!
Q15: WHY do I have to wear eyeglasses, dental braces, and wavy hair? Oy Vey!
Q16: WHY do I need a period for 33 years—once a month—for a week? Oy Vey!

Dearest Sharon Esther:

Bellum Omnium Contra Omnes
The War of **All** Against **All**

Thomas Hobbes

Dearest Thomas:

OMG: You can eat your own child for lunch!

Q: Who did you eat for lunch today?

THE 24-HOUR BLOODBATH
8 HOURS OF WORLD PEACE
WHEN EVERY LIVING BEING IS SLEEPING
16 HOURS OF DRAMA AND TRAUMA

LAW OF INEXTRICABILITY: PEACELESSNESS

Sharon Esther Lampert

PHILOSOPHER QUEEN SHARON ESTHER LAMPERT
LAWS OF INEXTRICABILITY
PEACELESSNESS
24/7
PREDATOR
AND PREY
LIFE
DEATH
24 HOUR CLOCK of DAYDREAMS & NIGHTMARES
EVERY DAY WE GET 8 HOURS OF WORLD PEACE
WHEN EVERY LIVING BEING IS SLEEPING
16 HOURS OF DRAMA AND TRAUMA

1. PEACELESSNESS

The Esoteric Law of the Inextricability of Life and Death

With history piling up so fast, almost every day
is the anniversary of something awful.
—Joe Brainard

LIFE & DEATH Are Inextricably Interwoven Into an Interdependent Matrix

24/7 PEACELESSNESS: There are millions of sentient life forms on the planet, and most fear each other, hate each other, and many eat each other to survive.

Predator & Prey: Each of us is both the predator and the prey.

Every day a bloodbath among living beings is in effect on the land, in the sea, and in the air, in the fight for resources to maintain biological survival.

The violence is indigenous; it knows no mercy, it has no compassion, it has no morality, it is brutal, it is painful, it is treacherous, and there is a continuous and constant suffering. Most predators eat their prey raw—no cooking required!

Cannibalism: You and your offspring are edible feasts for the right consumer—or you yourself can eat your own child for dinner to sustain your own life.

Life & Death are inextricably linked. If you want to live, you must consume the life of a simple or complex living being, or ingest a simple or complex living plant. Conversely, You cannot ingest and digest rotten animals or plants. Tragically, if you digest rotten animals or plants, you may die of food poisoning.

ALL EVIL IS JUSTIFIED: Throughout human history, we murder our own species in war games, but do not feed the enemy human remains to other starving animals or humans. We prefer to let enemy carcasses rot in unmarked mass graves.

During a war, millions of human lives are lost. There is no other sentient life form that murders its own kind in the name of religion or political ideology. Genocide (ethnic cleansing): humans are brutally tortured—then murdered.

In nature, dead animals do not waste away but are eaten by other animals on land, in the air, or in the sea. An airborn vulture will scoop down for a dead carcass. An alligator will exit the water to walk on land to consume a dead carcass.

We murder our own species—but our dead carcasses do not feed the hungry. **"Russian Deserter or Russian Fertilizer"** is a popular slogan of the war in Ukraine.

Dearest Sharon Esther:

"All truths are easy to understand
once they are discovered;
the point is to discover them."

Galileo di Vincenzo Bonaiuti de' Galilei

Dearest Galileo:

FIGHT TO LIVE
LIVE TO FIGHT
BORN TO DIE

LAW OF INEXTRICABILITY: LIFELESSNESS

Sharon Esther Lampert
PHILOSOPHER QUEEN

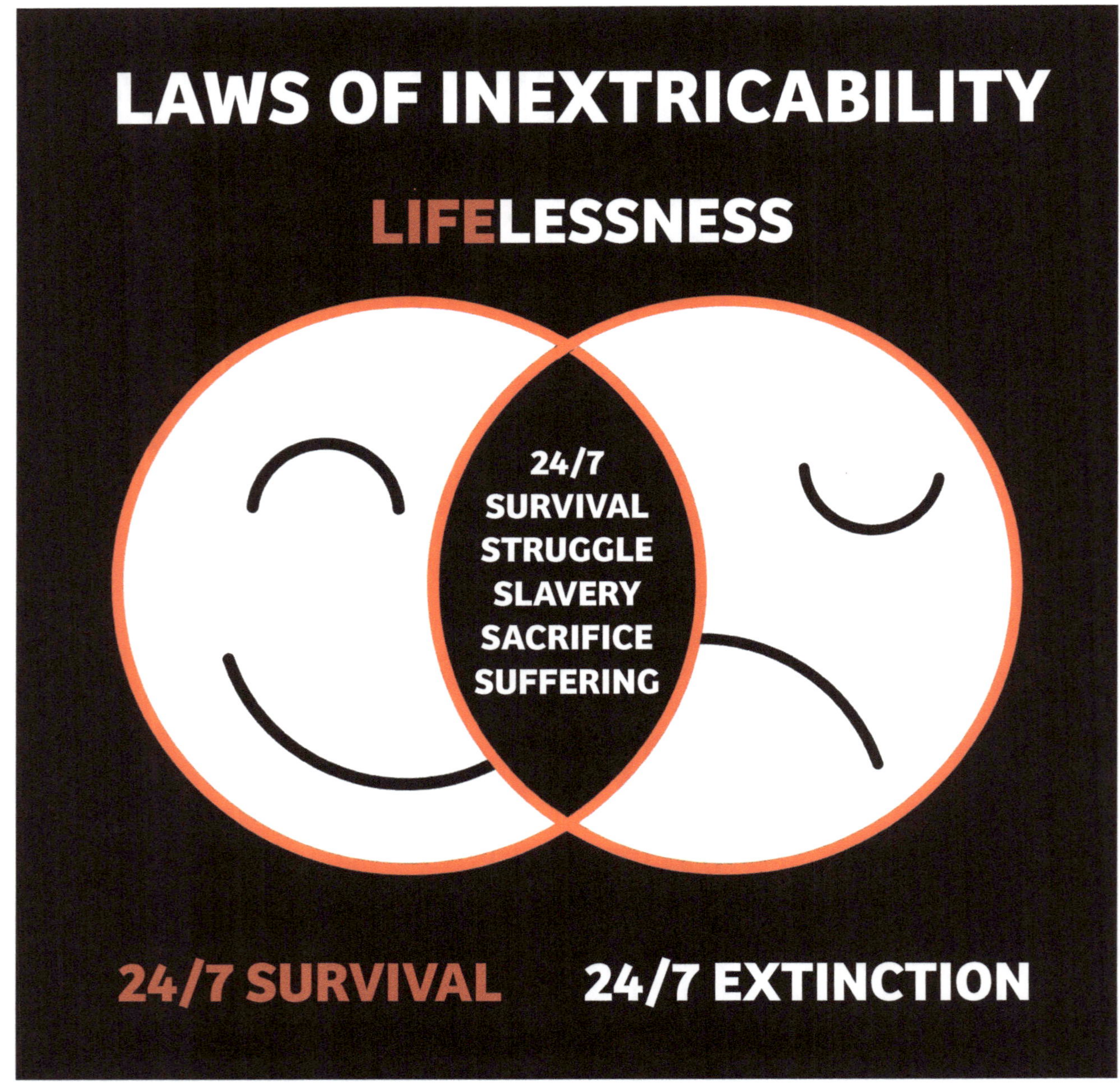

Read My Book: Temporary Insanity—Written in Letter S

2. LIFELESSNESS
The Esoteric Law of the Inextricability of Survival and Extinction

"It's only once that you accept that life
is a tragedy that you can start to live."
— Gloria Vanderbilt

SURVIVAL & **EXTINCTION** Are Inextricably Interwoven Into an Interdependent Matrix

"FIGHT TO LIVE, LIVE TO FIGHT, BORN TO DIE!" SEL

LIFELESSNESS: 24/7 We are forced to fight for our lives every hour of every day, from birth till death. Stop eating and drinking—and you will die!

LIFELESSNESS: 24/7 Your life will be extinguished if you don't fight for your life. It seems at first as if you are given a life, however, your life is not your own to do with it what you please. Paradoxically, we are asked to sacrifice our lives to maintain our biological survival—and this is the be all and end all of our existence.

24/7 LIFELESSNESS: SURVIVAL, STRUGGLE, SLAVERY, AND SACRIFICE
1. 24/7 Exertion & Recovery (1/3 of Life Is Sleeping)
2. 24/7 Energize: Eat 3X a day
3. 24/7 Defecation: Poop & Pee 3X a day
4. 24/7 Service to Society

We have to move and not move in accordance with nature. There is a delicate balance taking place between the energy required to move, and then once you have moved—the recovery time that takes place in sleep when you have to stop moving, so that you can start moving again (**LAW OF INEXTRICABILITY: EXERTION & RECOVERY**).

24/7: We are enslaved in a labor intensive, interdependent network of biological survival. As of 2025, there are 8-billion enslaved workers on the planet.

24/7: Our resources of food, clothing, and shelter have to be replenished, a perpetual state of enslavement. Like squirrels, we have learned how to store our survival food supply for months or years in freezers, storage bins, and basements.

"C'est La Mort"

Of all of our afflictions, the most equally distributed human suffering are the losses brought about by death.

We are all a witness to this very personal darkness, as one by one, every light among us, and everyone we love is taken from us.

We wait and watch, wondering how we will each meet our own devastating doom and final demise.

WHY the commotion when death's grip takes grasp; as if we must have done something wrong to deserve this final decree.

If we are being punished—then it is a collective punishment—leaving no soul among us spared. We are all living our lives on a dead-end street.

Life is Transient; Death is Permanent

Too many times, we are done with before we have completed our endeavors, "You have got to be kidding. This is the worst joke I have ever heard!"

Death is a humiliating end to a fervent and fanatical struggle to maintain one's life. Especially after all that time and effort to become educated, financially stable, secure and settled; or to finally reach a point when one may have something substantial to offer a family member or the next generation. After much toil—hard-earned **WISDOM** can only survive in books.

Tragically, every generation is left groping around in the vast lightlessness of their existence having to find their way from among the remains of written texts left by the finite few who have been able to create a flickering glint, gleam, or glimmer of **LIGHT** out of the rife darkness (**TOHU VAVOHU: GENESIS 1:2**)

In each and everyday that lives and dies, the **Angel of Death** is always standing beside us—in everything that we do—as death can overtake us at any time during our life: (1) treacherous weather; (2) toxic food; (3) diseases; (4) lethal virus; (5) fatal accident; (6) school shooter; (7) drowning in a pool; and (8) terrorist attack. The daily newspaper keeps tabs on the lucky and unlucky.

CURVEBALLS: Without warning, on land, sea or air, danger hovers and one's life can be revoked instantaneously (**LAW OF INEXTRICABILITY:** HOME**LESSNESS**)

As I was editing this book, the world reknowned vocalist Celine Dion was inflicted with a cruel rare medical disorder, Stiff-Person Syndrome, at the median age of 54. Celine is trapped by an illness of immense suffering with no cure. Brutal!

Recently, my dear friend Lenny Guttman was hospitalized with an emergency Septic-Shock infection that left him unable to move his limbs. Brutal! Thanks to antibiotics, he is recovering. Last week, he was walking with the help of a walker. He survived—a miracle!

Inevitably and eventually, each and every life is snuffed out or snuffs itself out like a flickering-melting candle. The **Angel of Death** is the final victor! Oy Vey!

Light is Momentary; Darkness Prevails: GENESIS 1:2

In our finite days, with every turn of the earth's axis, we sneak a peek at the new generation of people who may inhabit our homes and must water our lawns.

Well, however our individual cards play out, we did not exist for most of the history of the world and we will not exist for most of the world's inevitable future.

Life is Short; Death is Long

In summation, all the absolute truths, compiled and presented here for the very first time, ever in the history of our civilization, are aesthetically indigestible, however, they may prove to be the antidote to unveiling a greater connection to ourselves, in body, spirit, and mind.

Most certainly, we will develop a deepening, awakening consciousness that can enter the hearts of the living and propel civilization forward in search of a possible shared humanity of respect, compassion, and peacefulness.

Possible, however, not probable; the meaning of **LIFELESSNESS** is incomprehensible.

Death Is a Greater Force than Life—Life is Finite; Death is Final

WE FEAR LIFE & WE FEAR DEATH

NATURE IS THE CHURCH OF SATAN

Lars von Trier
FILMMAKER

Dearest Lars,

A 24-HOUR CLOCK OF DAYDREAMS & NIGHTMARES

FOR MANY PEOPLE DAYDREAMS ARE SCARIER THAN NIGHTMARES

CURVEBALLS: EXPECT THE UNEXPECTED
LAW OF INEXTRICABILITY: HOMELESSNESS

Sharon Esther Lampert
PHILOSOPHER QUEEN

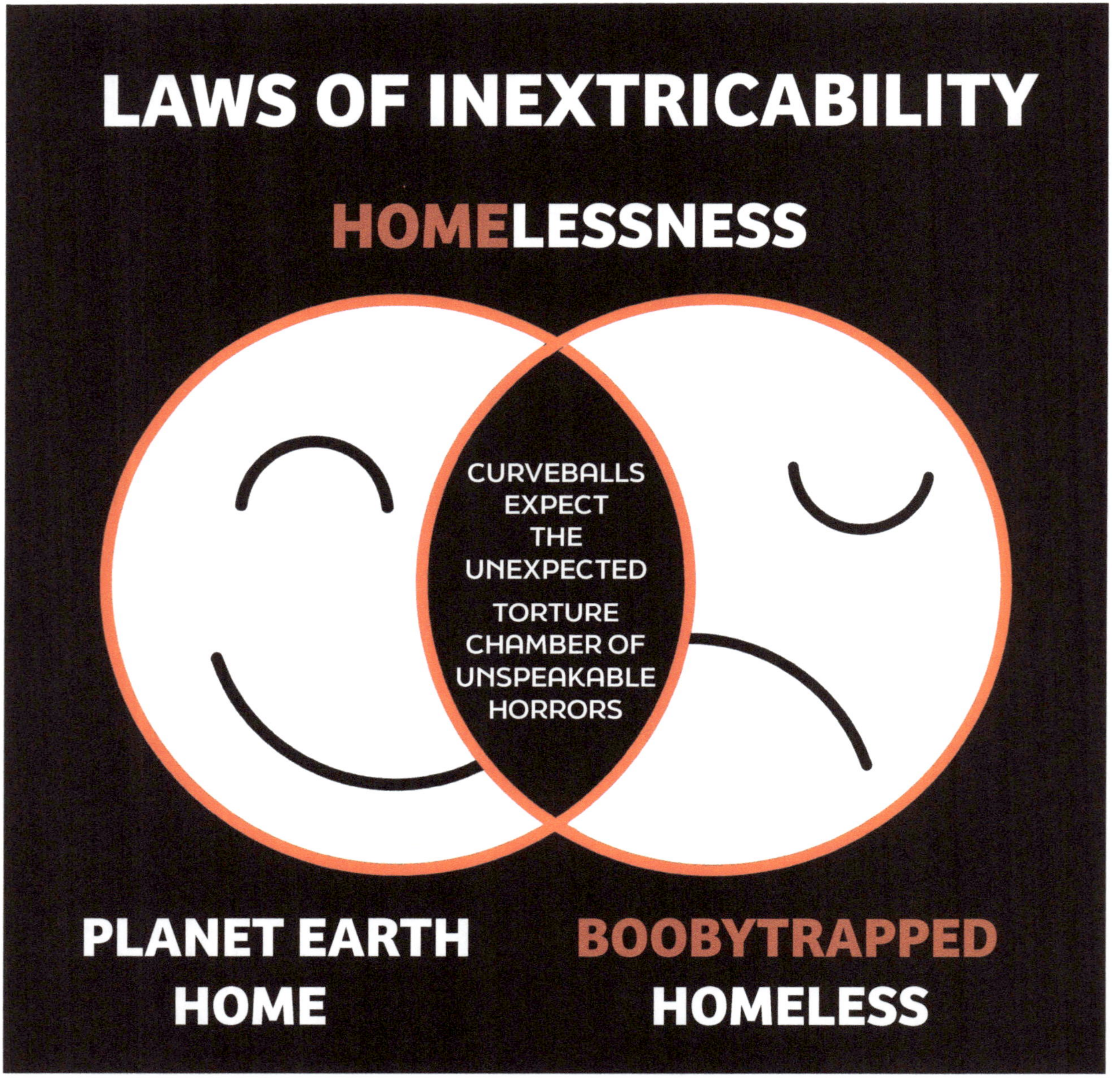

LAWS OF INEXTRICABILITY
HOMELESSNESS
CURVEBALLS
EXPECT
THE
UNEXPECTED
TORTURE
CHAMBER OF
UNSPEAKABLE
HORRORS
PLANET EARTH
HOME
BOOBYTRAPPED
HOMELESS

3. HOMELESSNESS

The Esoteric Law of the Inextricability of Security and Insecurity

CURVEBALLS: EXPECT THE UNEXPECTED

HOME & HOMELESSNESS Are Inextricably Interwoven Into an Interdependent Matrix

HOMELESSNESS: PLANET EARTH IS A BOOBY-TRAPPED TORTURE CHAMBER

Danger manifests everywhere! Q: What test did you take today?
HOMELESSNESS: If you make a wrong move—you will be doomed to death:
If you eat the wrong food—you may suffer food poisoning and die
If you swim in the ocean—you may drown
If you sit the hot sun—you may burn
If you sit in the cold—you may freeze
If you cross a street—you may get hit by a speeding car
If you fly on an airplane—it may crash
If you drive a car—you may have a fatal accident
If you make love to the wrong person—you may die from AIDS
If you are allergic to strawberries—you may die in an instant!
If you encounter a wild animal—you will be killed and eaten for lunch
If you encounter a human animal—you will be murdered and left to rot
If you are infected by an invisible-murdering-mutating microbe—you may die
and you will in turn infect and murder your loved ones—not your enemies

HOMELESSNESS: Nature's wrath is a daily occurrence that devastates and
destroys lives in less than a NY minute: earthquakes, hurricanes, tornadoes, floods,
blizzards, avalanches, wildfires, erupting volcanoes, and heat waves.
HOMELESSNESS: Unmitigated suffering from genetic diseases, and contagious
diseases wipe out millions of people in a heartbeat: AIDS.
24/7: We will never be able to feel **HOME-SWEET-HOME** on this planet: yesterday, today, tomorrow, and forevermore!

"When peace comes we will
perhaps in time be able to forgive
the Arabs for killing our sons —
but it will be harder for us to
forgive them for having
forced us to kill their sons.
Peace will come when the Arabs
will love their children
more than they hate us."

—Golda Meir, A Land of Our Own: An Oral Autobiography

Dearest Golda,

Point 1. **THE DARK TRIAD**
BORN IGNORANT, UNCONSCIOUS, IRRATIONAL

Point 2. **EVERY STORY IS THE SAME STORY: CAIN & ABEL**

Point 3. **EVIL: THE DOUBLE WHAMMY** (nasty nature & nasty nurture)

Point 4. THERE ARE 3 GAMES IN LIFE:

WIN! WIN! **WIN! LOSE!** **LOSE! LOSE!**
collaboration competition mutual destruction

Point 5. Centuries of Islamic **HOLY WARS!—ALLAHU AKBAR!**
Convert or Die!

Point 6. 100% of People Live Their Lives on a **DEFAULT** Setting

LAW OF INEXTRICABILITY: HUMANBEINGLESSNESS

Sharon Esther Lampert
PHILOSOPHER QUEEN

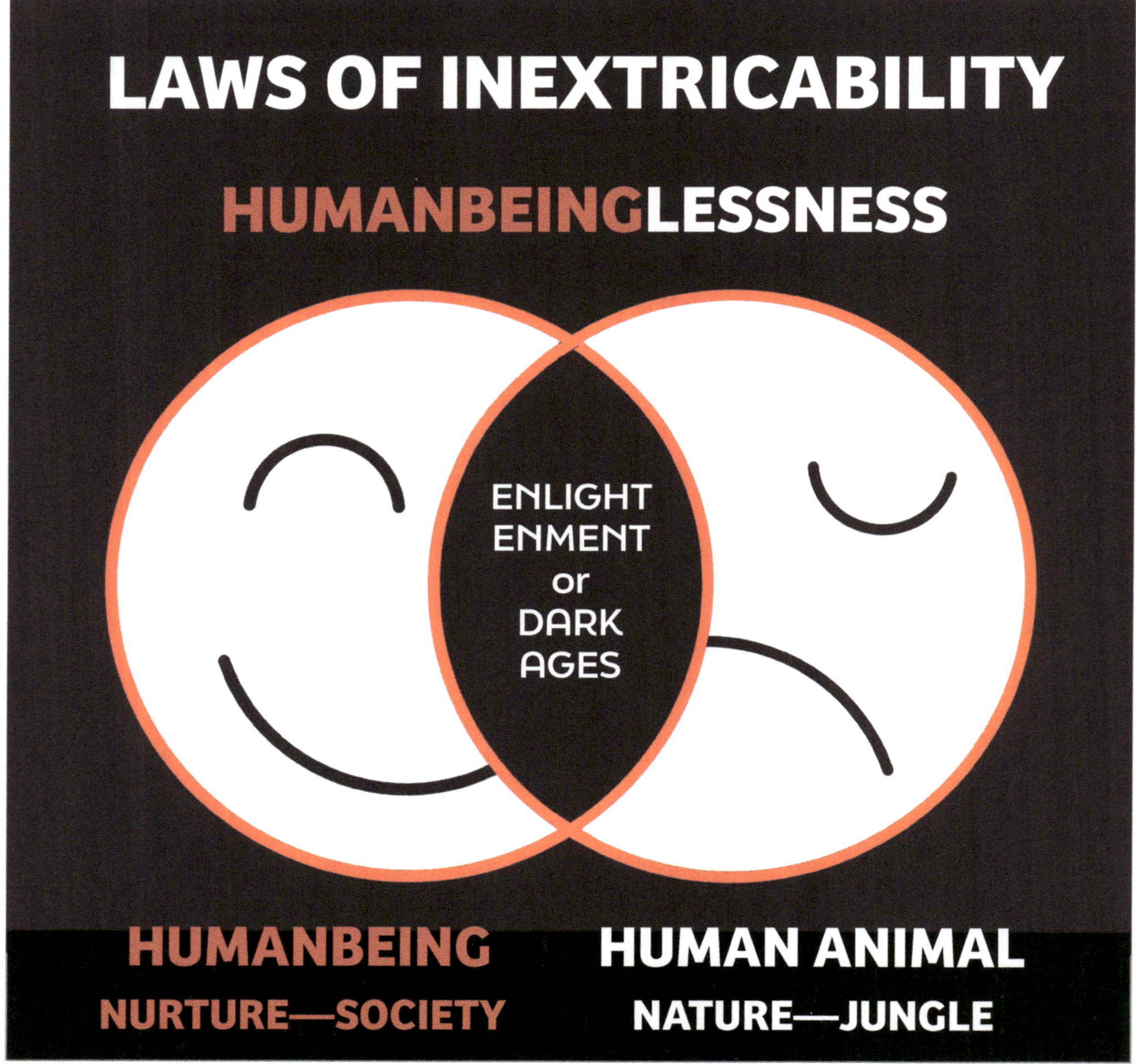

"Two Kinds of Men in This World:
Big Dicks Who Want to Build Something; and
small dicks who want to destroy everything."
Philosopher Queen Sharon Esther Lampert

4. HUMANBEINGLESSNESS

The Esoteric Law of the Inextricability of the Human Animal and Humanbeing

DARK AGES
When You Talk Like a Humanbeing and
Behave Like a Human Animal

ENLIGHTENMENT
When You Talk Like a Humanbeing and
Behave Like a Humanbeing

Sharon Esther Lampert
PHILOSOPHER QUEEN

The HUMAN ANIMAL and **HUMANBEING** Are Inextricably Interwoven
Into an Interdependent Matrix of NATURE & **NURTURE**

NATURE - HUMAN ANIMAL - LAW OF JUNGLE
NURTURE - HUMANBEING - CIVILIZATION

WILD ANIMALS ARE DOMESTICATED   WILD HUMAN ANIMALS ARE SOCIALIZED

The humanbeing is an "artificial construction," socialized to live in a
civilization by the triad of: LOVE, LEARNING, and LAWS.

ALL EVIL IS JUSTIFIED!
EVIL: "THE DOUBLE WHAMMY"
NASTY NATURE & NASTY NURTURE

YOU CANNOT NEGOTIATE WITH EVIL
DESTROY THE EVIL OR BE DESTROYED BY THE EVIL

PRINCESS KADIMAH: 8TH PROPHETESS OF ISRAEL

Human Animals Are Socialized and Schizophrenic

When you see a person behaving well, with social graces and manners, that person has been socialized, and is exhibiting artificially-contrived behaviors.

Every person has to navigate the difficult emotional terrain between the natural and innate animal instinct, and the artificially-contrived social behaviors set forth by the civilized society in which they live—and a SCHIZOPHRENIA sets in.

In the the public domain, people learn to behave with manners and social graces, and in the private domain, people continue to exhibit their natural and innate animal instincts.

Bar-Mitzvah Blues

During college, my part-time job was teaching children and adults to read Hebrew in preparation for their Bar & Bat-Mitzvahs (a photograph of me and my class was published in the book, "Jewish & Female" by Susan Weidman Schneider, 1984).

There is usually a 15-minute break for snacks, for cookies and juice. When the cookies are displayed on the table, one child, a boy, quickly runs to the plate of cookies, and fills his pocket with ten cookies, not thinking about whether there will be enough cookies for the other children.

When I approach him and explain the situation, that there are not enough cookies for the other children and, it is important to learn how to share, he asks me these questions:

Q: "What could possibly be good about sharing?'

"I now have ten cookies, and if I share—I will only have two cookies."

Q: "Why is sharing good?"

Q: "Why should I learn to share?"

As a Hebrew school teacher, I am supposed to have the right answer to these ethical questions, however, I am much more than a normal Hebrew-school teacher, I am also a **PHILOSOPHER QUEEN**; and I know that he is a human animal, and as a human animal, he is doing a great job—and I should applaud him with accolades:

1. He sees the cookies.
2. He runs fast.
3. He stuffs his pockets with ten cookies.

4. He has cookies for today and perhaps tomorrow.

5. He can trade extra cookies for services from friends.

He is following his natural and innate animal instincts of survival. I should applaud him, however, my job is to castigate him, and transform him into someone he is not—a humanbeing.

I am asking him to learn to behave in a way which is contrary to his natural and innate animal instincts—and a SCHIZOPHRENIA develops in him; the schizophrenia between his natural and innate animal instinct, and his learned socialized behaviors that are required of him to live in a civilization.

He has to learn to live in an artificially contrived environment that is often contrary to his natural animal instincts. This is precisely where his emotional problems begin and end, and for that matter, this is also the crux of the matter, for most of the other emotional problems of every person who is living within a society.

It is difficult to maintain the delicate balance because from the personal to the private, because you have two personas that are in opposition to each other.

For instance, in a civilized Hebrew school society, the artificially contrived rule is that every child will get one cookie, and if there are enough cookies for all of the children, then each child will get two cookies, or if there are only five cookies left on the plate, I may raffle them off to the child who knows the answer to a Jewish trivia question.

African Farmers vs. Elephants

NATURE, a weekly television show on channel 13, had a program on the wild elephants in Africa. The elephants venture into the fields of the farmers, and eat their crops during the night. Farmers would kill the invading elephants. LOSE! LOSE! **GAMEPLAN** (mutual destruction).

The only way to stop the elephants was to capture them, domesticate them, and use them for transportation.

The farmers would then be in charge of feeding them, and the elephants no longer ravaged the farmers crops.

People and elephants were working in unison, exchanging needs in order to survive and coexist. In sum: Socialization of human animals and domestication of wild animals working in unison is a WIN! WIN! **GAMEPLAN** (collaboration).

Dearest Sharon Esther:

"A man does what he must—
in spite of personal consequences,
in spite of obstacles and dangers,
and pressures—and that is the
basis of all human morality."

President John F. Kennedy

Dearest President John F. Kennedy:

Q: How do you live a moral life in an immoral universe?

8 UNPREDICTABLE OUTCOMES

You Can Do It All Wrong and Get It All Right!
You Can Do It All Right and Get It All Wrong!

LAW OF INEXTRICABILITY: MORALITYLESSNESS

Sharon Esther Lampert
PHILOSOPHER QUEEN

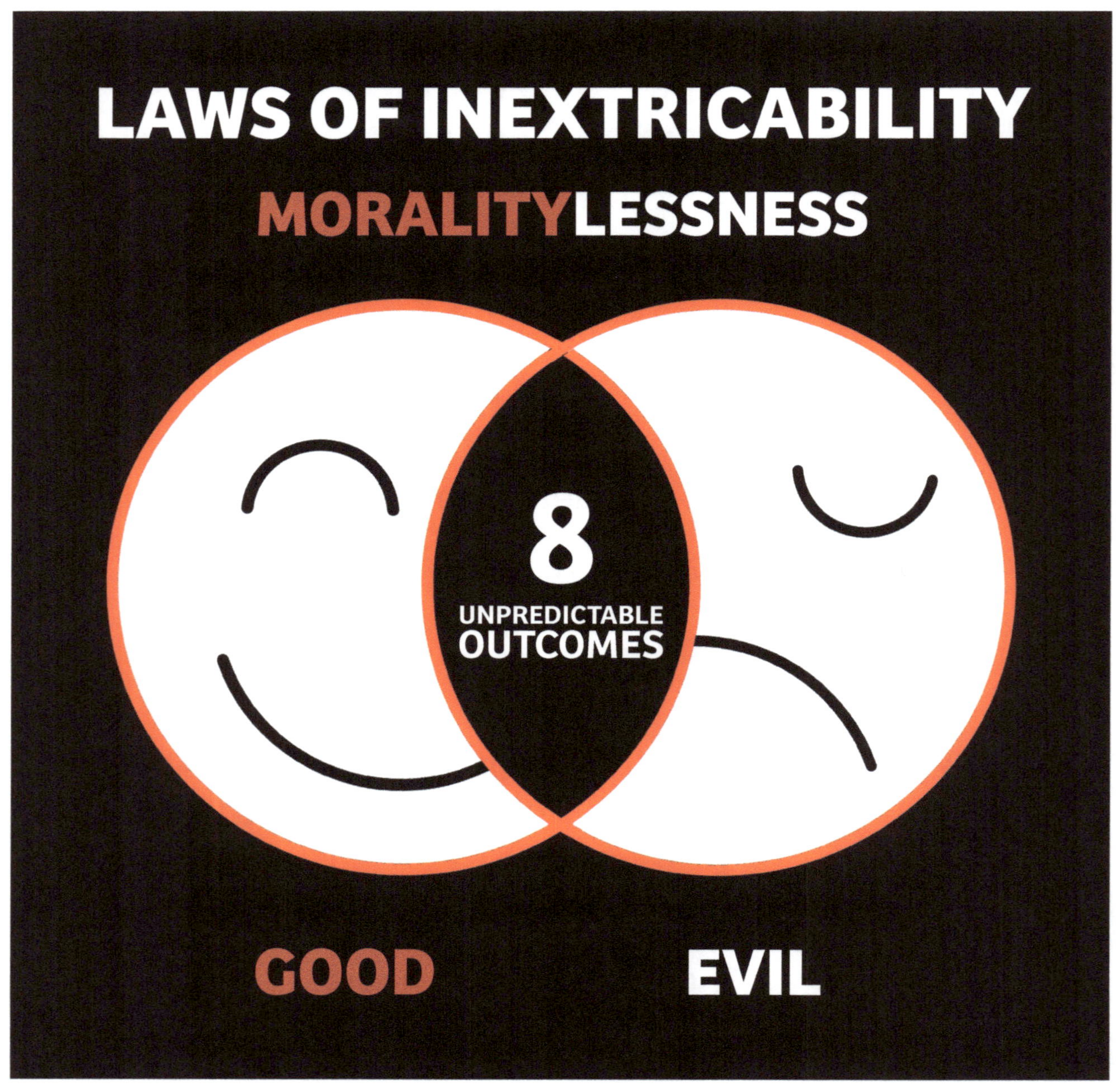
LAWS OF INEXTRICABILITY
MORALITYLESSNESS
8
UNPREDICTABLE
OUTCOMES
GOOD
EVIL

5. MORALITYLESSNESS

The Esoteric Law of the Inextricability of Good and Evil

"You Can Do It All Right and Get It All Wrong, or
You Can Do It All Wrong, and Get It All Right!"

Sharon Esther Lampert
PHILOSOPHER QUEEN

Q: How do you live a moral life in an immoral universe?

GOOD & EVIL Are Inextricably Interwoven Into an Interdependent Matrix

MORALITYLESSNESS:
In an immoral world, **GOOD & EVIL** have **EIGHT UNPREDICTABLE OUTCOMES:**

1. Good turns into good
2. Good turns into good & evil
3. Good turns into evil
4. Good turns into nothing
5. Evil turns into evil
6. Evil turns into good & evil
7. Evil turns into a good
8. Evil turns into nothing

Q: Why do bad things happen to good people?

MORALITYLESSNESS
8 UNPREDICTABLE OUTCOMES

For Instance:
MALALA: EVIL LEADS TO A GREATER GOOD
When Malala was shot in the head by a Taliban terrorist in Pakistan, she was airlifted to the United Kingdom—and reaped many rewards and awards:
- She earned a first-rate education in the United Kingdom
- She was honored with a Nobel Prize
- She raises money to educate women in third-world countries

For Instance:
MARTHA SMART: GOOD LEADS TO EVIL AND EVIL LEADS TO A GREATER GOOD
Martha Smart's father gave a homeless man a job, and the man abducted his 14-year-old daughter, and drugged and raped her every day for nine months.
- Martha Smart is an international advocate on behalf of abducted children

Dearest Sharon Esther:

"Life can only be
understood backwards,
but it must be lived forwards."

Søren Kierkegaard

Dearest Søren:

Point 1.
From conception to death, you undergo the process of METAMORPHOSIS

Point 2.
Everyone gets a different set of cards in life.
What you do with the cards you are dealt in life determines your DESTINY!

Point 3.
Q: Are you living your life by DEFAULT or by DESIGN?
100% of People Live Their Lives on a DEFAULT Setting!

LAW OF INEXTRICABILITY: CHOICELESSNESS

Sharon Esther Lampert
PHILOSOPHER QUEEN

Read My Book: SPERM MANIFESTO: THERE IS ONLY ROOM FOR ONE AT THE TOP

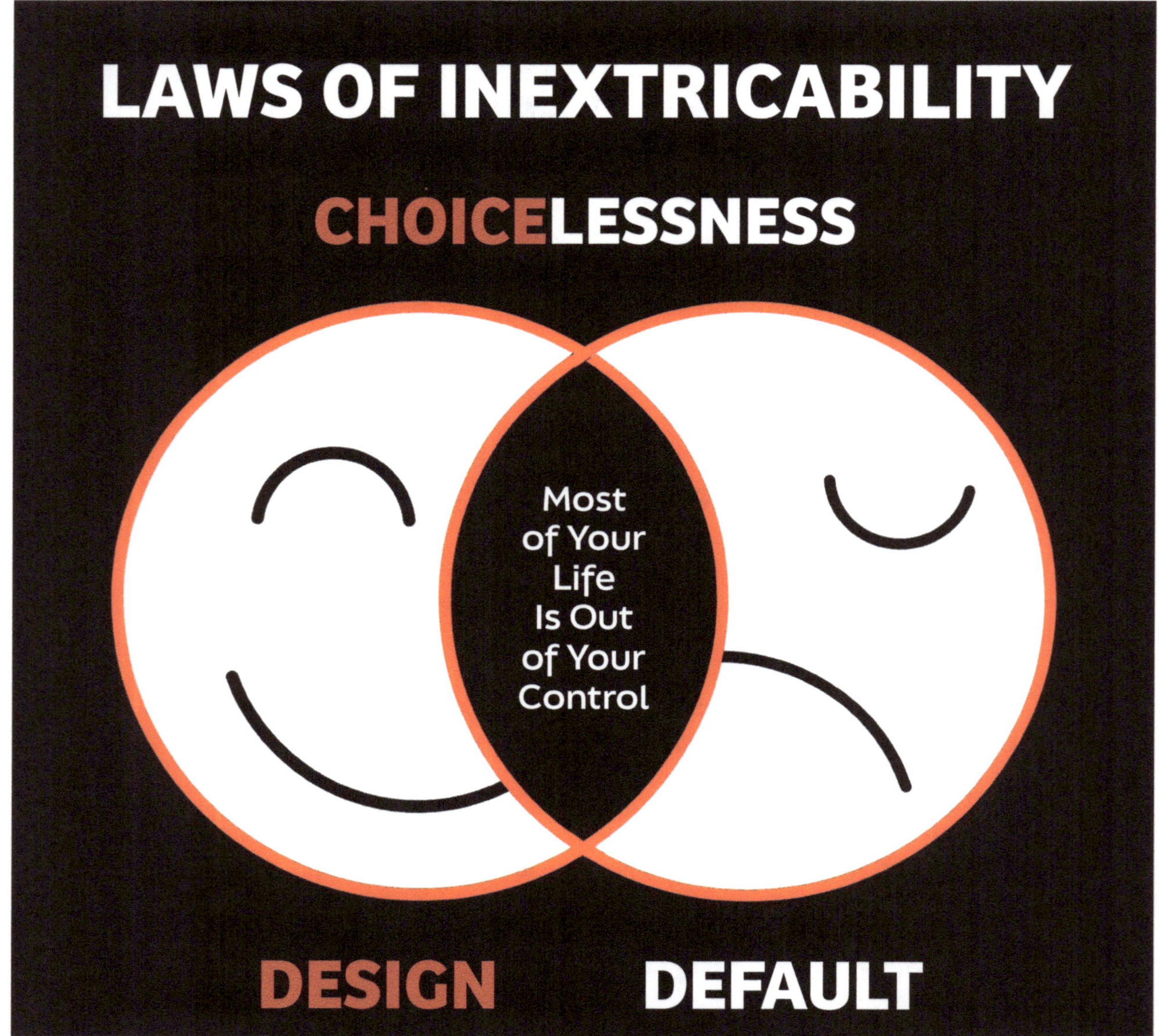

"Everyone Gets a Different Set of Cards in Life —
What You Do with the Cards You Are Dealt in Life
Determines Your Destiny!"

Philosopher Queen Sharon Esther Lampert

6. CHOICELESSNESS
The Esoteric Law of the Inextricability of Default and Design

Living Your Life By **DEFAULT** and **DESIGN** Are Inextricably Interwoven Into an Interdependent Matrix

You have to play the cards you are dealt in life; what you do with the cards you are dealt in life determines your **DESTINY**...

CHOICELESSNESS: You have no say over the most important decisions in your life. Vital decisions were made for you without your consent or consideration—**MOST OF YOUR LIFE IS OUT OF YOUR CONTROL!**

1. When you are born and when you will die
2. Your DNA: Height, Eye Color, Body Type, Gifts and Talents
3. Your Parents
4. Your Religion
5. Your Nationality
6. Native Language
7. Your Early Education and Social Conditioning
8. Your Allies & Enemies
9. Life's Vicissitudes: **EXPECT THE UNEXPECTED: CURVEBALLS**

DEFAULT: Birth- Age 18

100% of people live their lives by **DEFAULT.** We are born into authoritarian birth bubbles. Our parents dictate our lives and are our first role models. On one hand, your **DEFAULT** setting may have been a blessing—not a curse. On the other hand, if your disagree with your programming, it may be construed as a rebellious act. Your personal affinities will be dismissed if they conflict with the norms of your culture. Any deviation from the norms of your culture will be frowned upon—and you may even be excommunicated and abandoned like a stray dog on the street.

DESIGN: Ages 18+

Once you reach the legal age of 18, you can set yourself free from your **DEFAULT** setting, and begin the journey of making choices that take into account your personal preferences and perogatives. You will always bear the imprint of your **DEFAULT** setting, even as you make course corrections and live your life by **DESIGN.**

Dearest Sharon Esther:

"There is always some
madness in love.
But there is also always
some reason in madness."

Friedrich Nietzsche

Dearest Friedrich:

Q1: What is Love?

Q2: Why is the person you and married the same person you hate and divorced?

LAW OF INEXTRICABILITY: LOVELESSNESS

Sharon Esther Lampert
PHILOSOPHER QUEEN

Read My 4 Books on Love:

1. The Philosophy of Love: If You Practice SELF-LOVE You Will Never Spend a Day in Therapy
2. LOVE YOU MORE THAN YESTERDAY: 14 Relationship Strategies for HAPPILY EVER AFTER
3. CUPID —Written in Letter C, The Awesome Art of Alliteration Using One Letter of the Alphabet
4. SEX ON A PLATE: FOOD AS FOREPLAY: THE COOKBOOK OF EVERLASTING LOVE

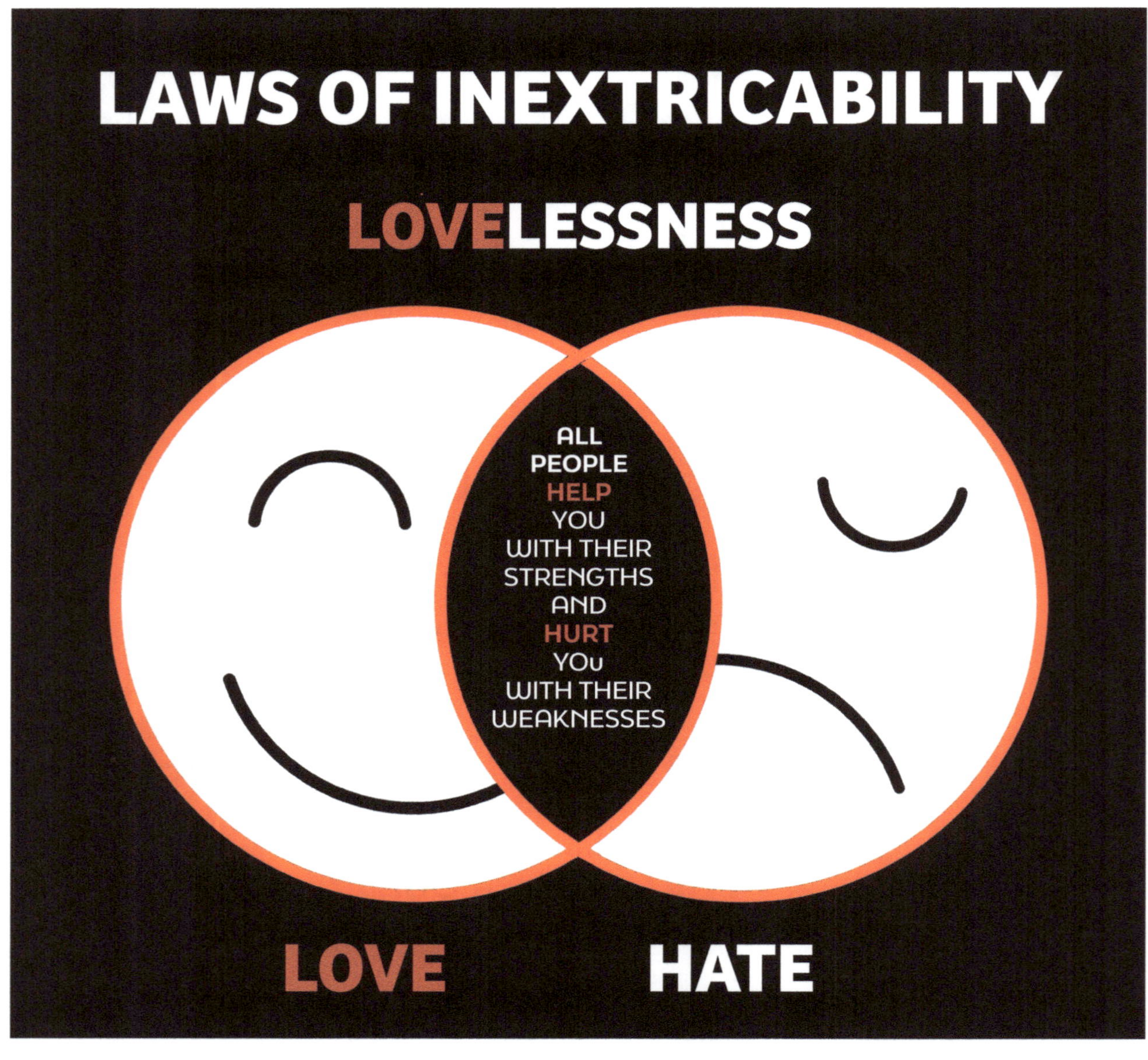

100% of Relationships Are Love-Hate Relationships

7. LOVELESSNESS
The Esoteric Law of the Inextricability of Love and Hate

LOVE & HATE Are Inextricably Interwoven Into an Interdependent Matrix.

LOVELESSNESS
All People **HELP** You with Their Strengths and **HURT** You with Their Weaknesses!

Part 1. Q: What Is Love?
        A: You Don't Find Love, You Create Love: Respect, Kindness, and Empathy
        A: Love Is a Daily Practice

Part 2. There Are Two Kinds of Relationships: **ME & WE**
    ME: The Most Imortant Relationship Is the One You Have with Yourself
    ME: TRUE LOVE IS **SELF-LOVE**

    WE: You Can Never Know Another Person—Love Is Blind!
    WE: Love From Outside Yourself Is **BONUS LOVE**
    WE: LAW OF INEXTRICABILITY: **LOVE**LESSNESS: All People **Help** You
        with Their **Strengths** and **Hurt** You with Their **Weaknesses**
        100% of Relationships Are Love-Hate Relationships

Part 3. There Are Two Kinds of Love:
    1% True Unconditional Love and 99% Conditional Transactional Love

    1. 1% **TRUE LOVE**: Real But Rare—Like Winning a Lottery Ticket
       • The Loss of **TRUE LOVE** Is Excruciating; Time Cannot Heal the Loss
       • **TRUE LOVE** Is Till Death Do You Part

    2. 99% Conditional Transactional Love
       • They Love Only What They Want From You
       • Relationships Have an Expiration Date

Dearest Sharon Esther:

"In the land of the blind,
the one-eyed man is king."

Desiderius Erasmus

Dearest Desiderius:

Q: Do you remember your
9 months in the womb?

On your death bed, you
will not be able to remember
having lived a life?

The most important question
every generation should ask Is:

Q: HOW SCARY IS THAT?
THE DARK TRIAD

WE ARE BORN INTO WORLD UNCONSCIOUS, IGNORANT, AND IRRATIONAL
WE ARE BORN THIS WAY! WE WILL LIVE THIS WAY! WE WILL DIE THIS WAY!

IN EVERY GENERATION THE BLIND LEAD THE BLIND

TRIAL'N'ERROR: EVERYTHING GOES WRONG BEFORE ANYTHING GOES RIGHT!

WE FEAR LIFE AND DEATH!

LAW OF INEXTRICABILITY: MINDLESSNESS

Sharon Esther Lampert
PHILOSOPHER QUEEN

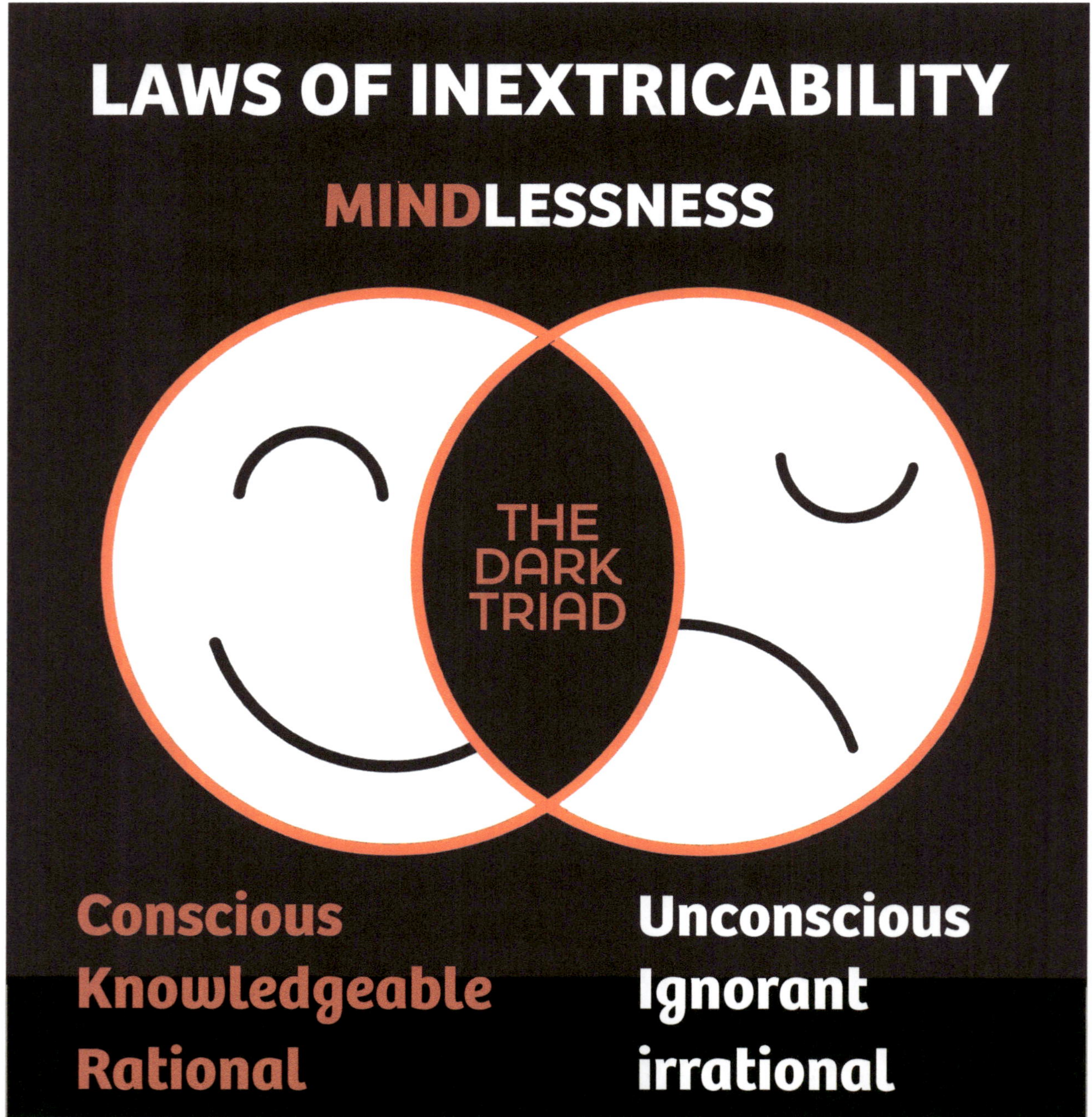

EVERYTHING GOES WRONG, BEFORE ANYTHING GOES RIGHT! Philosopher Queen Sharon Esther Lampert

8. MINDLESSNESS

The Esoteric Law of the Inextricability of Unconscious and Conscious

UNCONSCIOUS and CONSCIOUS/ IRRATIONAL and RATIONAL
Are Inextricably Interwoven Into an Interdependent Matrix

MINDLESSNESS: THE DARK TRIAD

1. We are born into the world: UNCONSCIOUS, IGNORANT, and IRRATIONAL
2. IN EVERY GENERATION THE BLIND LEAD THE BLIND
3. Everything Goes Wrong Before Anything Goes Right!
4. 99% of People Wish They Could Live Their Lives Over Again to Get It Right

MINDLESSNESS: The state of consciousness is not possible—only an approximation—or a matter of degree of consciousness is attainable.

We have faulty memories and must learn and relearn whatever we are taught, time and again. In old age, debilitation, senility, and dementia rob us of our minds. On our deathbed—we cannot remember having lived a life.

Education, therapy, and meditation are some of the means available to achieve a degree of consciousness, yet these modalities have inherent limitations.

We live in a quadruple fog: mind fog, body fog, world fog, and **WHY** fog:

Mind Fog:
The first fog is the fog that exists inside of our own minds. Our ideas, thoughts, and feelings are invisible and intangible entities. **WE FEAR LIFE AND DEATH!**

Body Fog:
The second fog is the fog that exists inside of our bodies. People live inside of bodies that they do not understand. Our bodies are on autopilot. We do not turn our bodies on or turn them off. We must eat and drink 3X a day to keep our bodies turned on.

World Fog:
The third fog is that we don't understand the world that exists around us. Millions of sentient life forms inhabit our world—most are living their lives on autopilot: eat, pee, poop, clean, work, and sleep.

The WHY? of Existence: The Ever Impenetrable Fourth Fog

It is hard for me to fathom the fact that billions of people lived and died on Planet Earth, before the day I arrived, and that the world is still engaged in bloody-battlefield wars and genocides that span weeks, months, and years. People across the globe experience famine, poverty, and suffer the indignity of **IGNORANCE** due to the lack of an education.

The world we inhabit operates according to dynamics that are beyond our understanding, and beyond our control.

We don't flip a switch and turn the sun, moon or stars on or off.

Our lives are a precarious happening in history. We don't ask to be born, however, we are forced here, and thrust into circumstances that are not of our choice.

Most of our lives will come and go very quickly, and are of no earth shattering consequence.

Most of us will soon be forgotten by our closest loved ones, until of course, reflective glimpses of our physical or spiritual selves appear within their reflective mirrors.

We will live longer life spans than our cherished pets, but shorter life spans than many species of trees.

In light of what we are able to comprehend, artificial fictions abound in every facet of our lives.

Each person is besieged with having to qualify every statement that is heard with: **Q: "Is it true?"** or **Q: "Is it false?"**

Our minds ask these questions every day with the hope of attaining a discernible glimmer of a comprehensible picture of an objective truth that exists outside of our subjective minds.

We are unceasingly engaged with making sense out of our perplexing, puzzling, and paradoxical existence.

How many people know their own minds, or understand how the bodies that they occupy function, or comprehend the world around them—even though we have an intimate experience of our own minds, inhabit our own bodies, and interact with the world, day in and day out.

Perhaps, it is explainable by the understanding that we have had to ingest, digest, and absorb many artificial fictions about ourselves, and about our existence; and this blinding fogginess, a white fog, leaves each and every generation, living in a state of chaos, conflict, and confusion.

We forge ahead, with invention and innovation in science, medicine, computers, technology and space travel; but more often, we rage ahead, with daily civil strife.

When our savage temper boils over, we embroil ourselves in blood-curdling genocide. Acts of genocide cross all civilizations and all continents: Armenians, European Jews, Cambodians, Native Americans, Nanking Massacres of the Chinese, Tibetans, Rwandans, Bosnians, Kosovo Albanians, and East Timorians.

A case can be made, that we lack a capacity to decipher the true state of our existence—and make light of it. As a result, our artificial fictions make the most sense that is possible, in a given time and place in history.

The choices we made were based on the artificial fictions we believed in at that time. We pay a steep price as we become part of the imperfect problem, instead of becoming part of the imperfect solution. We have had no other choice, as we were unable to choose, having ingested and digested so many artificial fictions that were poisonous to our mind, heart, and soul.

Our decisions are colored by extreme emotional states, blinding rages of pure hatred for ourselves, and of the other.

We are figuratively and literally willing to pay an arm and a leg as well as give an arm and a leg for more guns, more bombs, more hand grenades and more land mines.

If we had to grade ourselves, our report card would read, an A in effort in manufacturing weapons of mass destruction: nuclear, biological, and chemical warfare, and a D in self-love, and an F in self-control.

The other must die for one reason or another, and
we are willing to sacrifice ourselves to get the job done!

In lieu of what we have been able to accomplish to dignify our lives through science, medicine, and technology—we continue to cheapen our lives through the desecration of war and genocide.

The devastation of ourselves and of the other makes the most sense.

Simply, **MIGHT = RIGHT**. All other resolutions become unintelligible.

In a war, we are willing to have ourselves killed to defend our artificial fictions. The killing of millions is acceptable in a state of war.

When we go to war, however, death and destruction of thousands becomes our only resolve. **WAR** is a **LOSE! LOSE Gameplan** (mutual destruction). The objective is to put the other in the cemetery. Billions of dollars of resources are squandered.

ALL EVIL IS JUSTIFIED!

PHILOSOPHER QUEEN SHARON ESTHER LAMPERT

GOD OF WHAT?

Measure the Pleasure
Life is a Gift

Measure the Pain!
Life is a Punishment

15 Esoteric Laws of Inextricability

ONLY GOD KNOWS WHY? AND GOD KNOWS BEST!

Is Life a Gift or a Punishment?

15 Absolute Truths

THE END OF FEEL GOOD PSYCHOBABBLE TO NUMB THE PAIN AND KEEP THEM SANE!

 It is time to make international laws that state that **WAR** is illegal, and that governments must do everything in their power to prevent them.

Our soldiers should be put out to pasture, and reassigned to life-sustaining objectives.

How is it that we forget that it is our beloved children that we send to fight our wars for us.

What gives us the right to sacrifice our beloved children in the prime of their lives to solve problems? This is institutionalized child abuse.

Every nation across the globe creates an indestructible army of destructible soldiers. Both sides of a battlefield will be dismembered limb by limb, before all limbs become completely severed, only then, will it be deemed a completely reasonable and rational choice to surrender with the one remaining limb, and raise a white flag.

What in the whole wide world could possibly be worth all that effort? And if a country can strive to produce such a seemingly effortless task of destruction, why can't the same massive effort be used to promote life?

After the gut-wrenching drama of war unfolds, the soldiers who survive are plagued by incurable traumas.

Our victorious soldiers return home and become our broken children.

Many future generations are damaged from the effects of war on previous generations.

After the dead have been buried, we will all agree that war is the wrong choice, that tomorrow is a new day and change is possible, and that this was the last war, however, in each and every generation, the infernos of hatred are inextinguishable, and the resulting bonfires of war engulf us once again.

War is a billion-dollar profitable enterprise; education is a non-profit enterprise.

Our artificial fictions manifest themselves everywhere, at any given time and place.

Moreover, most of our own personal lifetime experiences fade fast into oblivion. Our human memory is a subjective experience that is faulty and fleeting.

Our only recourse is the written word. The written word helps us keep track of where we have been, where we are, and where we are going; and is our only means of remembering the WHAT? WHO? WHEN? and WHERE? of our experience.

The WHY? of our existence is the ever impenetrable fourth fog.

"Everywhere I go a poet has been there before me."

Sigmund Freud

DEADICATION

Every single second of the day,
A suffering is taking place:
Shake-me. Wake-me. Save me.

Every single second of the day,
My belief in a God is shaken:
Shake-me. Wake-me. Save me.

Death is a greater force than life,
Doctors can't heal death:
Shake-me. Wake-me. Save me.

Evil is a greater force than good;
Judges can't punish evil:
Shake-me. Wake-me. Save me.

Good deeds are fragile glass;
Evil deeds are solid rock:
Shake-me. Wake-me. Save me.

Joy dissipates; Pain endures.
Shake-me. Wake-me. Save me.

GOD IS DEAD

Poet Sharon Esther Lampert

#1 Poetry Website for Student Projects
The Greatest Poems Ever Written on Extraordinary World Events

POETRY WORLD RECORD "Through The Eyes of Eve"
120 Words of Rhyme from One Family of Rhyme

Note:
1. This is the first copy of my poem.
2. The poem was expanded into a book entitled:
DEADICATION

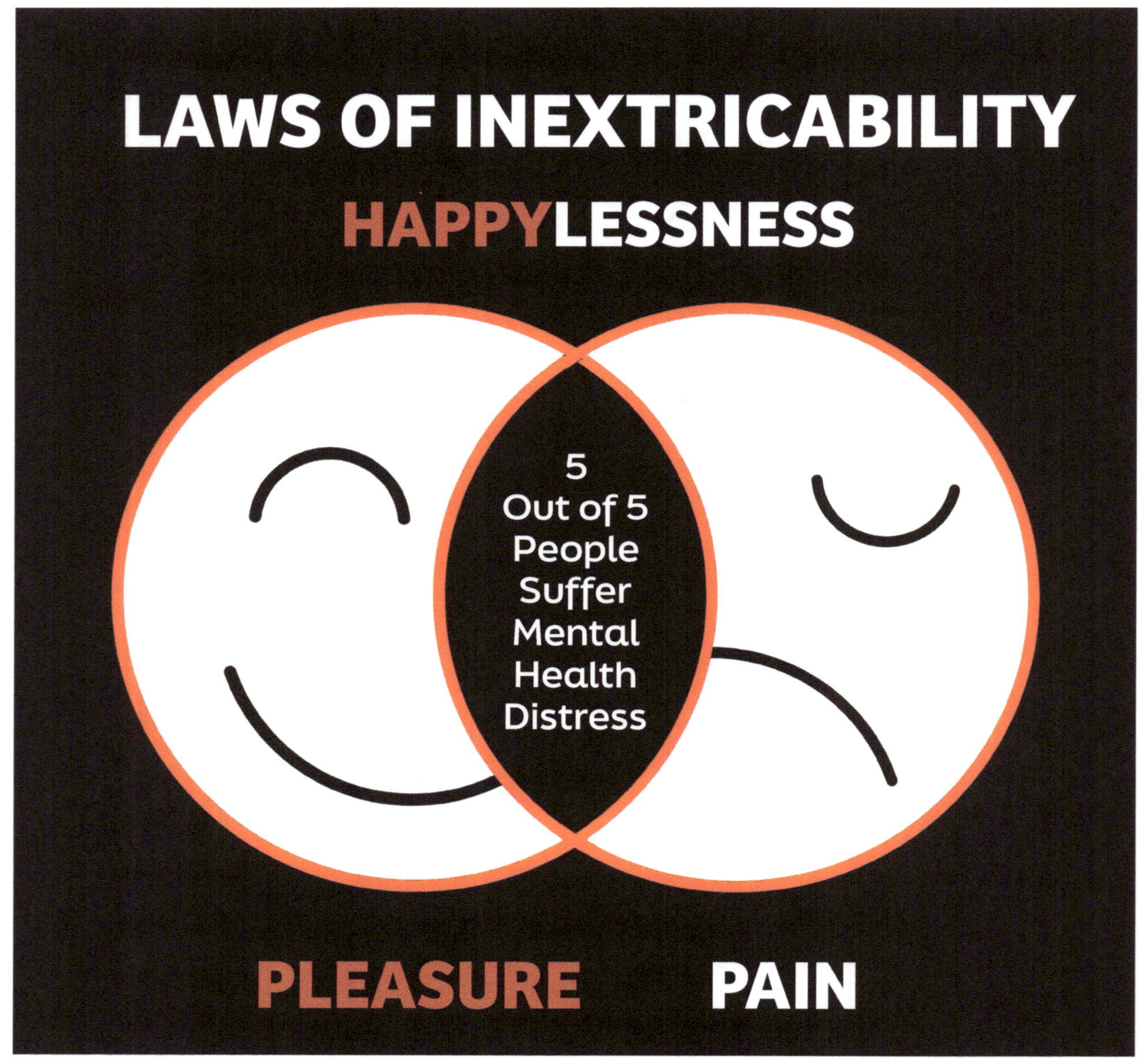

Read My Book: **MONSTERS:** Broken World of Broken People

PHILOSOPHER QUEEN Sharon Esther Lampert

9. HAPPYLESSNESS

The Esoteric Law of the Inextricability of Pleasure and Pain

In her diary, just days before her death,
Frieda Kahlo penned her final entry stating:
"I hope the exit is joyful—and I hope never to return."

— Frieda Kahlo

PLEASURE & PAIN Are Inextricably Interwoven Into an Interdependent Matrix

HAPPYLESSNESS: E.g., Cake brings us pleasure & getting fat from cake brings us pain.

HAPPYLESSNESS: For instance, one of the greatest joys of life is the blessing of giving birth to a healthy baby—even if the birth of the baby entails excruciating pain to the mother.

HAPPYLESSNESS: For instance, we comfort a crying baby who is in distress because of hunger and a wet diaper. First day of life: the pain of hunger, the pleasure of breast milk, and the fight for survival: **LIFE**LESSNESS.

HAPPYLESSNESS: For instance, as the crawling baby undergoes metamorphosis into a walking toddler, the pain & pleasure of falling and walking sets the stage for the continuous undulating state of the pain & pleasure of failure and success that are inextricably integral to the undulating peaks & pitfalls of life's daily vicissitudes.

HAPPYLESSNESS and **PEACE**LESSNESS: For instance, you will inflict pain on a living animal to kill it to eat it—and simultaneously, the ingestion of the animal will bring you pleasure. **One animal's pain is another animal's pleasure.**

HAPPYLESSNESS IN SPORTS:

For instance, every sport played has a winner and a loser. At the final buzzer, the winning team is celebrating with a bottle of champagne, while the losing team is devastated, depressed, and dejected. **WIN! LOSE! GAMEPLAN** (competition)

HAPPYLESSNESS: SCHOOL SHOOTING MASSACRE

For instance, in the school shooting in Parkland Florida, 17 students were brutally murdered, and 17 other students were brutally shot and will have to live out their lives with bullet fragments lodged deep inside of their bodies. They simultaneously feel the pleasure of survival & the pain of their indelible wounds.

Dearest Sharon Esther:

"Sometimes it's an advantage to be disadvantaged."

Novel: "Only Time Will Tell" by Jeffrey Archer

Dearest Jeffrey:

LAW OF INEXTRICABILITY: MERCYLESSNESS
YOUR DISADVANTAGE IS SOMEONE'S ADVANTAGE

Sharon Esther Lampert
PHILOSOPHER QUEEN

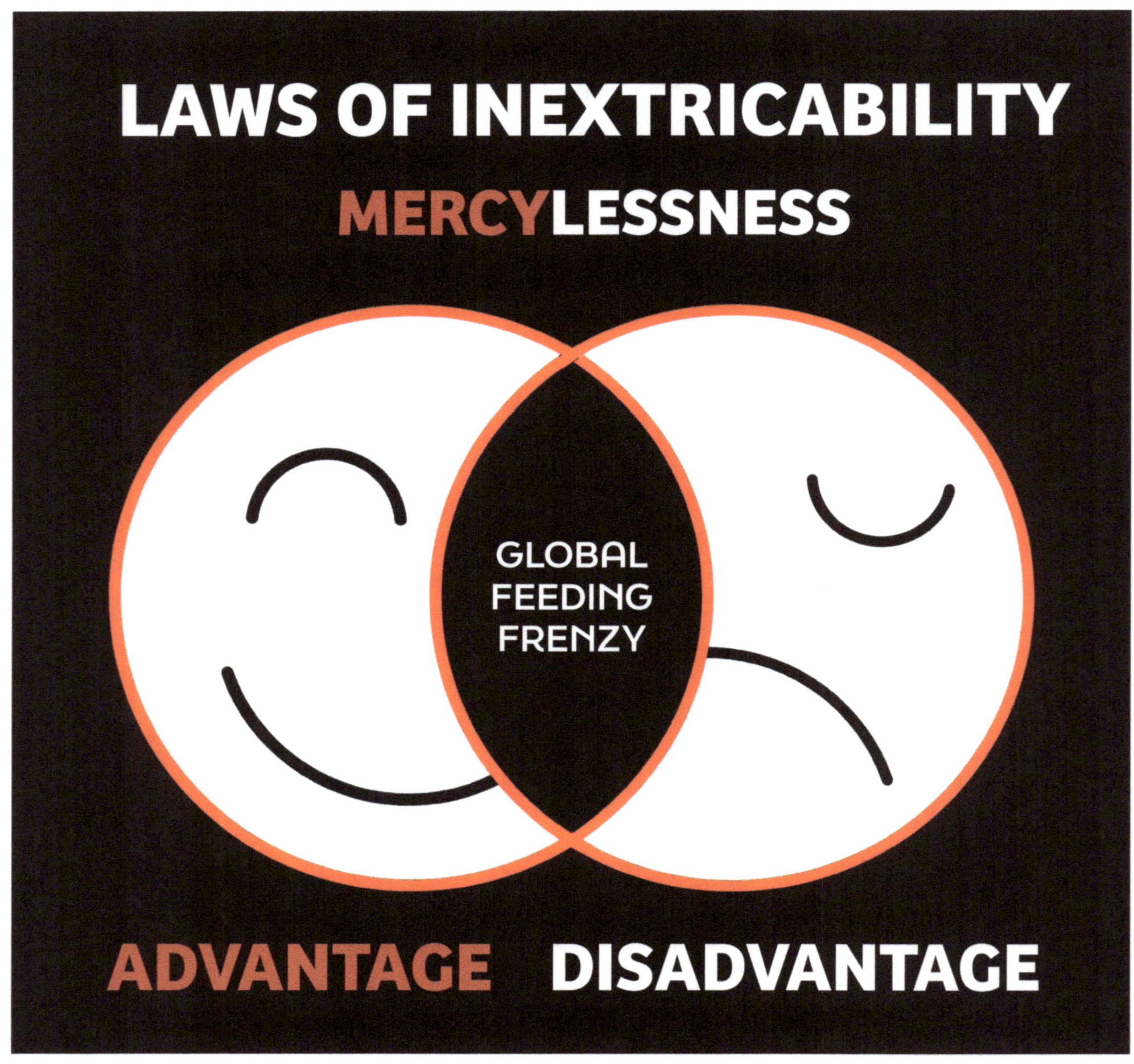
LAWS OF INEXTRICABILITY
MERCYLESSNESS
GLOBAL FEEDING FRENZY
ADVANTAGE DISADVANTAGE

10. MERCYLESSNESS

The Esoteric Law of the Inextricability of Advantage and Disadvantage

MERCYLESSNESS: ADVANTAGE and DISADVANTAGE
Are Inextricably Interwoven Into an Interdependent Matrix

"A Jewish woman had two chickens.
One got sick, so the woman made chicken soup
out of the other one to help the sick one get well.
Henny Youngman

MERCYLESSNESS: For every **DISADVANTAGE**, there is someone who will gain an **ADVANTAGE** and benefit from your **DISADVANTAGE.** You are either on the side of **ADVANTAGE** or you on the side of **DISADVANTAGE: A GLOBAL FEEDING FRENZY!**

HUNGER: For instance, you are hungry, and want to eat something. You are hungry, so you are on the side of the **DISADVANTAGE.** You have many choices: You can grow the food, you can purchase the food in a supermarket, or you can make a restaurant reservation! The supermarket and restaurant are on the side of the **ADVANTAGE.** We are usually hungry at least three times a day, a recurrent state of perpetual **DISADVANTAGE.** We wake up every day hungry and disadvantaged.

ILLNESS: For instance, you don't feel well. You call a doctor and make an appointment. You are ill and on the side of the **DISADVANTAGE,** and the doctor is on the side of the **ADVANTAGE.**

IGNORANCE: For instance, you are ignorant and have to get an education. You are on the side of the **DISADVANTAGE,** and the teacher and school are on the side of the **ADVANTAGE.**

SEXISM: For instance, you are born female, and the male gender disinfranchises you as being of lesser value than being born an advantageous male.

RACISM: For instance, you are the minority—and the majority disinfranchises you as being of lesser value than the advantageous majority.

SPORTS: For instance, in football, you foul a player on the opposing team, and the opposing team is awarded a penalty kick, scores a goal, and wins the game. Your **DISADVANTAGE** is their **ADVANTAGE** and ultimately their victory.

Invisible **Mutating-Murdering-Monster Microbes**

Dearest Sharon Esther:

"Viruses are indeed exquisitely deadly,
but they have provided the world with
some of its most important innovations.
CREATION and **DESTRUCTION** join together once more."

Carl Zimmer
SCIENCE WRITER

Dearest Carl:

We Are Building Our Lives on a Sand Trap!

LAW OF INEXTRICABILITY: CREATIONLESSNESS

Sharon Esther Lampert
PHILOSOPHER QUEEN

Read My Book:
TEMPORARY INSANITY: We Are Building Our Lives on a Sand Trap —Written in Letter S
Awesome Art of Alliteration Using One Letter of the Alphabet (6 Books)

LAWS OF INEXTRICABILITY
CREATIONLESSNESS
Everything
Has an
Expiration
Date
CREATION
DESTRUCTION

11. CREATIONLESSNESS

The Esoteric Law of the Inextricability of Creation and Destruction

CREATION and DESTRUCTION Are Inextricably Interwoven Into an Interdependent Matrix

CREATIONLESSNESS: Everything we create will undergo: decay, decomposition, downgrade, decline, destruction, deterioration, disease, and death.

CREATIONLESSNESS: We have to replenish all of our resources. Our food supply undergoes decay, our clothes undergo deterioration, our car is downgraded the day after we purchase it. Our computers have a 3-year shelf life.

CREATIONLESSNESS: We create life by sexual intercourse—even though we know that all life created will be doomed to decay, disease, and death.

CREATIONLESSNESS: For instance, you plant a lemon tree, and gather your ripe lemons. Lemons are nutritious and packed with vitamin C. Lemons can be stored at room temperature for a week, or stored in the refrigerator for a 4-6 weeks, or stored in the freezer for 3-4 months. The same lemon that boosted your intake of vitamin C, and helped you ward off the common cold—is the same expired lemon that will undergo decay and decomposition, and will may make you sick from food poisoning, and may even kill you!

CREATIONLESSNESS: For instance, when your immune system is robust, you are healthy and your body is able to fight off invaders: bacteria, viruses, and toxins. When your immune system is weak, you will catch a cold, a flu, or a virus. As you age, your health will decline, and your immune system's efficiency will decrease and be downgraded—until the finality of disease, senility, and death.

CREATIONLESSNESS: In wars, we willfully drop billion-dollar bombs to destroy our enemies. In minutes, thousands of years of a civilization is destroyed. We turn our own children into soldiers—who kill and will be killed.

CREATIONLESSNESS: We are born ignorant, irrational, and unconscious. Any toddler can throw a drinking glass on the floor, and smash it into smithereens of shards of glass that cannot be glued back together—but the toddler needs an education to create a new drinking glass.

Death Is a Greater Force Than Life—Life is Finite; Death Is Final.

Dearest Sharon Esther,

"Philosophy is the science which considers truth."

Aristotle

Dearest Aristotle,

THERE IS ONLY ONE
TRUTH
NO ONE HAS THE TRUTH

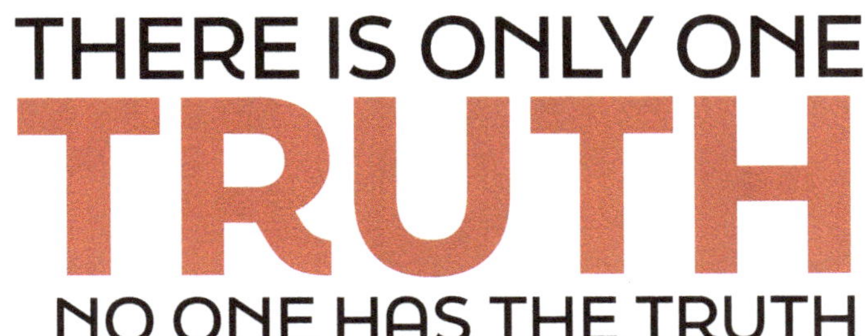

LAW OF INEXTRICABILITY: TRUTHLESSNESS

Sharon Esther Lampert
PHILOSOPHER QUEEN

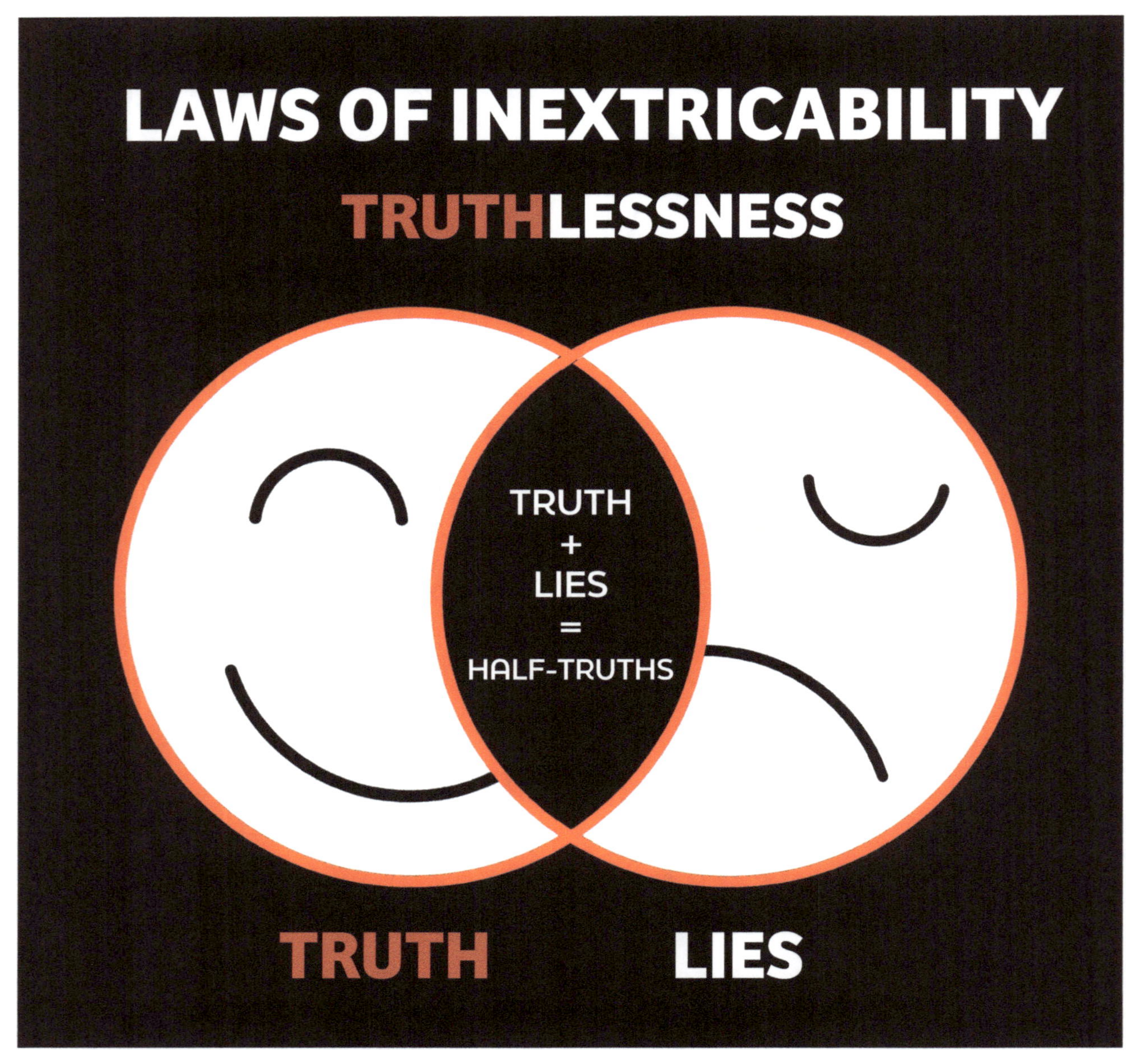

LAWS OF INEXTRICABILITY
TRUTHLESSNESS
TRUTH
+
LIES
=
HALF-TRUTHS
TRUTH
LIES

12. TRUTHLESSNESS

The Esoteric Law of the Inextricability of Truth and Lies

TRUTH + LIES Are Inextricably Interwoven
Into an Interdependent Matrix of **HALF-TRUTHS**

TRUTH is not self-evident; it undergoes distillation and purification.

You cannot make an ugly girl beautiful—and you cannot educate an imbecile—who lacks the capacity for intelligence.

1. To discern truth from lies or facts from fiction, you need THINKING tools. THINKING is a learned skill; it requires an education:

    a. Critical Thinking Tools
    b. Mathematical Thinking Tools
    c. Scientific Thinking Tools
    d. Creative Thinking Tools (10 Laws of Genius & Creativity)

2. **CRITICAL THINKING TOOLS:** Learn how to sift through primary sources to discriminate between opinions, belief systems, and facts.

3. The attainment of **WISDOM** is the highest priority: Transform Information into knowledge and knowledge into **WISDOM**.

4. **MISINFORMATION** is false or inaccurate information—getting the facts wrong.

5. **DISINFORMATION** is false information which is deliberately intended to mislead—intentionally misstating the facts.

6. **Daily Practice:**
    1. Q: Fact or Fiction?
    2. Q: Truth or Lie?
    3. Q: Wisdom: Light or Darkness?

"Beware When the Great God
Lets Loose a Thinker on this Planet."

Ralph Waldo Emerson

Dearest Ralph:

Q: Why is it NORMAL
for billions of people to talk to GOD
— but CRAZY if GOD talks to me?

LAW OF INEXTRICABILITY: GODLESSNESS

Sharon Esther Lampert
PHILOSOPHER QUEEN

Read My Book: WHO KNEW GOD WAS SUCH A CHATTERBOX

Dearest Sharon Esther,

"I want to know God's thoughts—the rest are details."

— Albert Einstein

Dearest Albert,

God Is Not Physics—The Laws of the Universe—God Is Metaphysics!
God Is an Invisible and Intangible Entity Like Your Mind, Thoughts, and Ideas
—Beyond the Scope of Scientific Inquiry.

GOD IS GO! DO!

GOD CAN ONLY DO FOR YOU
WHAT GOD CAN DO THROUGH YOU!

PRAY As If It All Depends on **GOD!** **WORK** As If It All Depends on **YOU!**

Dear God, I am naked!	GOD IS GO! DO!	Sew Clothes
Dear God, I am hungry!	GOD IS GO! DO!	Plant a Seed
Dear God, I am thirsty!	GOD IS GO! DO!	Purify Water
Dear God, I want to fly like a bird!	GOD IS GO! DO!	Build an Airplane
Dear God, I want to swim like a fish!	GOD IS GO! DO!	Build a Ship
Dear God, I want to visit the moon!	GOD IS GO! DO!	Build a Spaceship

LAW OF INEXTRICABILITY: GODLESSNESS

Prophet Sharon Esther Lampert

LAWS OF INEXTRICABILITY
GODLESSNESS
TWO
WORLDS
PHYSICAL
and
META
PHYSICAL
Metaphysical World
GOD IS GO! DO!
10 LAWS OF GENIUS AND CREATIVITY
Physical World
LAWS OF INEXTRICABILITY
Boobytrapped
Torture Chamber of
Unspeakable Horrors
GENESIS 1:2 TOHU VAVOHU

13. GODLESSNESS

The Esoteric Law of the Inextricability of Physical and Metaphysical Worlds

GODLESSNESS: The Physical World and the Metaphysical World
Are Inextricably Interwoven Into an Interdependent Matrix

There are Two Inextricable Worlds: Physical World and Metaphysical World

Physical World:

1. Physical World Is Organized By: LAWS OF INEXTRICABILITY
2. Physical World is a booby-trapped torture chamber of unspeakable horrors
3. The physical world is organized by the **SPERM MANIFESTO: THERE IS ONLY ONE ROOM FOR ONE AT THE TOP.** The drama that occurs inside the womb is the same drama that occurs outside the womb! The **GOOD, GREAT, and GIFTED** leave a legacy for generations—all future generations will stand on their shoulders

Metaphysical World: My Books Write Themselves!

1. Metaphysical World Is Organized By: 10 ESOTERIC LAWS OF GENIUS AND CREATIVITY
2. **Metaphysical World:** Your mind, thoughts, and ideas are invisible and intangible entities—beyond the scope of scientific inquiry
3. God Is Not Physics—The Laws of the Universe—God Is Metaphysics
4. GOD IS GO! DO!
5. GOD Can Only Do For You What GOD Can Do Through You!
6. Pray As If All Depends On GOD; Work As If It All Depends On YOU!

For instance: Metaphysical World: Gold Ring and Green Heart

At the age of 9, I lost a gold ring with a peridot heart on the beach. At the age of 18, I was taking a walk on the same beach—and found a gold ring with a peridot heart in the sand.

I wear my metaphysical gold ring peridot heart 24/7

For instance: Metaphysical World: Cat Plays a Piano (schmaltzy.com)

On the way home, I see a child's toy piano in the garbage. I decide to take it home, wash it, and give it to my cat **SCHMALTZY.** He folds his paws under his body and plays the piano like a virtuoso (YouTube video). **SCHMALTZY** becomes a **WORLD FAMOUS PIANO-PLAYING CAT.**

For instance: Metaphysical World: Favorite Perfume

I listed my favorite perfume on a notepad—a dear friend gave me a bottle as an unexpected gift!

"The hour of departure has arrived
and we go our ways.
I to die and you to live.
Which is better?
Only GOD Knows!"

SOCRATES
PHILOSOPHER KING

Dear **GOD**,
Q: Do you even exist?

And **GOD** said:
Q: Do you even exist?

HERE TODAY—GONE TOMORROW!

SHARON ESTHER LAMPERT
PRINCESS KADIMAH: 8TH PROPHETESS OF ISRAEL

Read My Book: WHO KNEW GOD WAS SUCH A CHATTERBOX

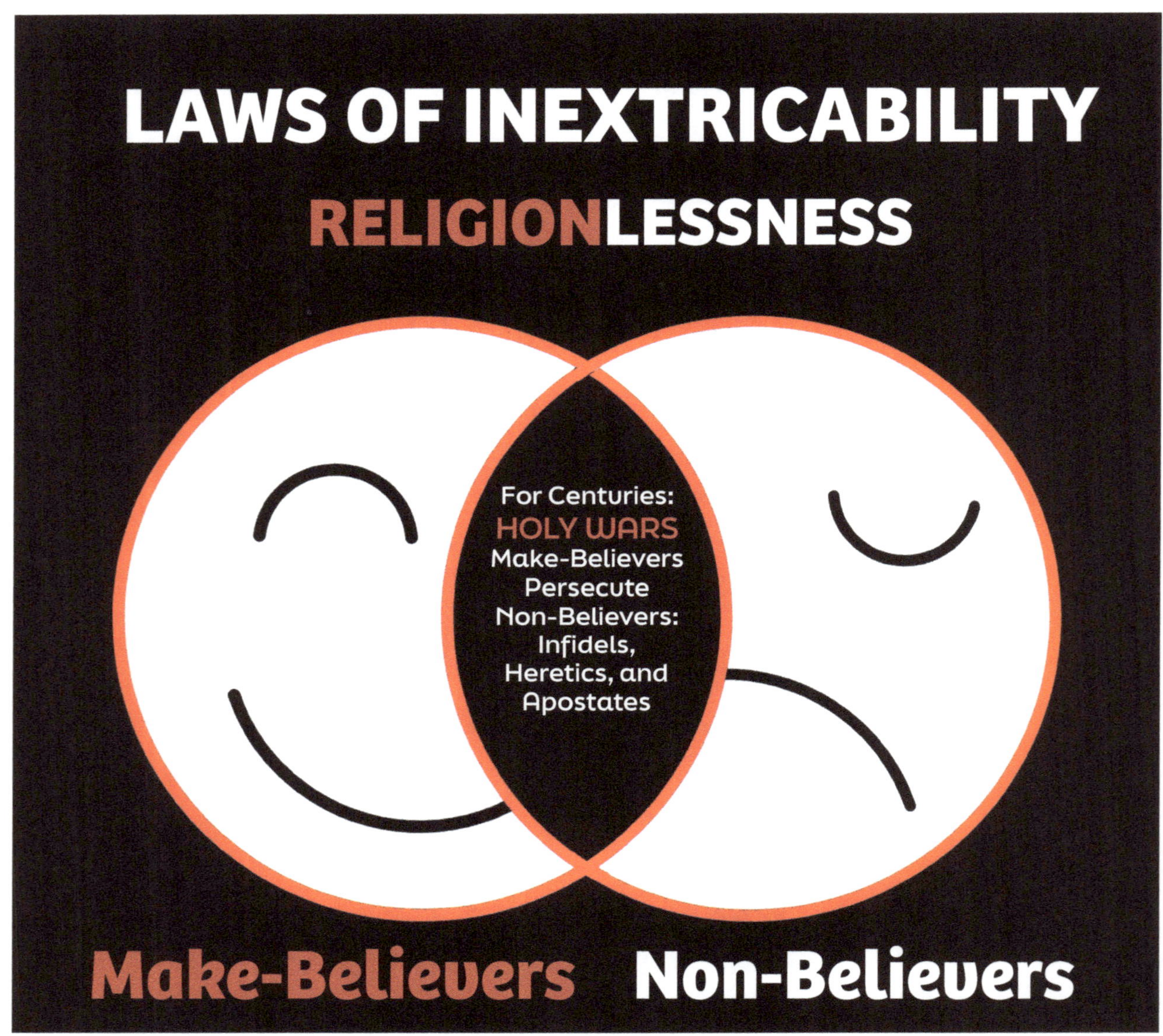

"THERE ARE NO BELIEVERS!
THERE ARE ONLY MAKE-BELIEVERS AND NON-BELIEVERS"
Philosopher Queen Sharon Esther Lampert

14. RELIGIONLESSNESS

The Esoteric Law of the Inextricability of Make-Believers and Non-Believers

RELIGIONLESSNESS: Make-Believers and Non-Believers Are Inextricably Interwoven Into an Interdependent Matrix

"Generally speaking, the errors in religion are dangerous; those in philosophy only ridiculous."

Philosopher David Hume

THERE ARE NO BELIEVERS!
THERE ARE ONLY MAKE-BELIEVERS AND NON-BELIEVERS

Philosopher Queen Sharon Esther Lampert

There are more than 4000 religions. There are thousands of holy books. Muslims pray to GOD 5X a day. Jews pray to GOD 3X a day. Hindus 2X or 3X a day. There are religious holidays every week:

Q: Where do the make-believers find the time in their overscheduled holiday planners for centuries of religious persecution of the non-believers: infidels, apostates, and heretics?

1. Centuries of **HOLY WARS** in the name of a GOD: **Convert or Die!**
2. A **ZERO TOLERANCE POLICY** for alternative spiritual paths: **Burn in Hell!**
3. Islamic **JIHADISTS** dominate the headlines: They massacre, rape, behead, loot, and force conversions (e.g., Nigeria and Sudan).

4000 Religions Recite One Prayer:

"Dear God, Help Me, Heal Me, and Protect Me from Harm": RELIGIONLESSNESS!"

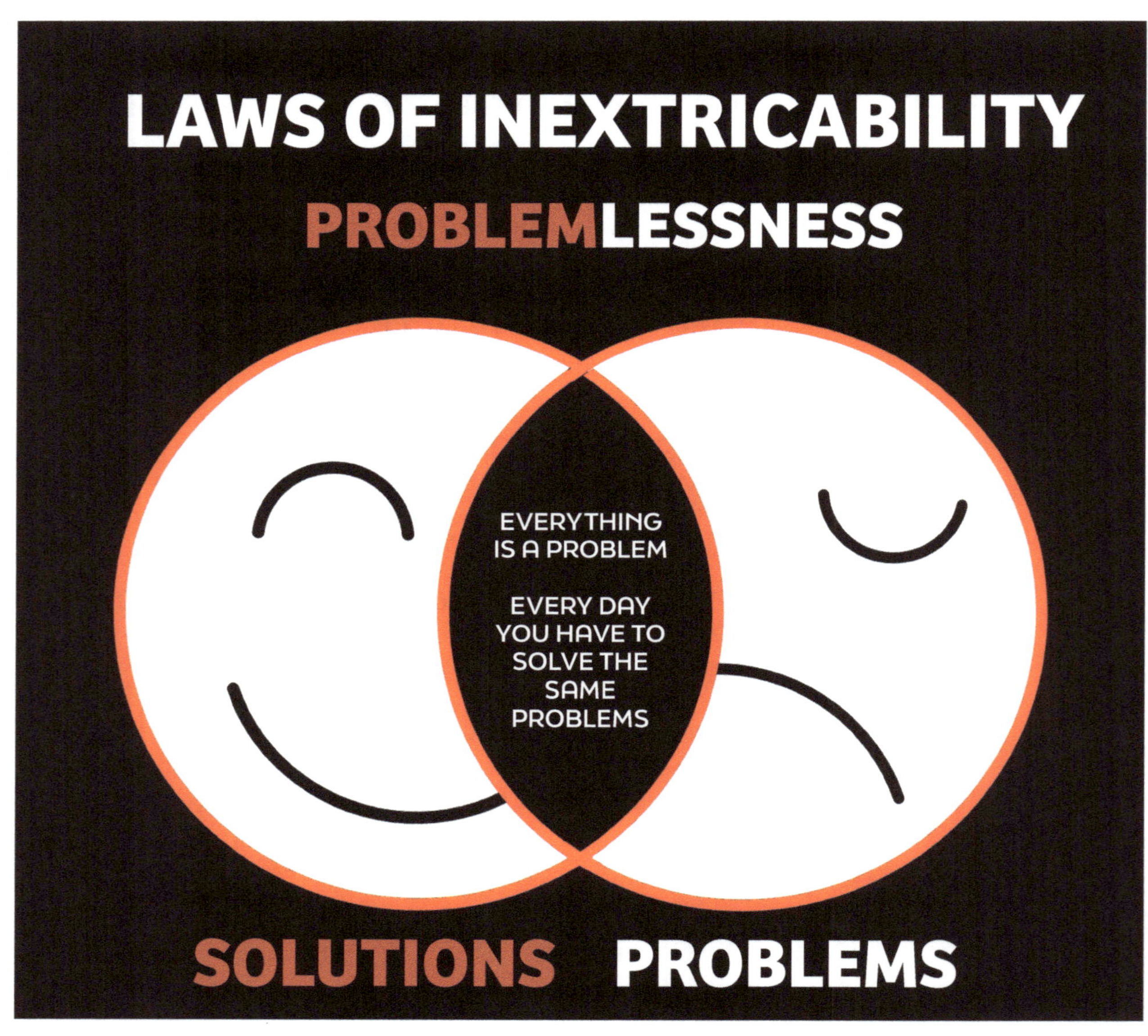
LAWS OF INEXTRICABILITY
PROBLEMLESSNESS
EVERYTHING IS A PROBLEM
EVERY DAY YOU HAVE TO SOLVE THE SAME PROBLEMS
SOLUTIONS   PROBLEMS

15. PROBLEMLESSNESS

The Esoteric Law of the Inextricability of Problems and Solutions

PROBLEMLESSNESS: Problems and Solutions Are Inextricably Interwoven Into an Interdependent Matrix

EVERYTHING IS A PROBLEM IN NEED OF A SOLUTION

Everyday Problems:
I'm hungry, I'm thirsty, and I'm naked
I'm born into the DARK TRIAD: Ignorant, Unconscious, and Irrational

Everyday Solutions:
I need food and water
I need a school and an education
I need a home and a bed to sleep
I need a toilet and toilet paper
I have a monthly period and I need a napkin
I need to check the weather forecast for rain, snow, or sun

The problems listed above cannot be eradicated. Every day of your life, you have to solve the same problems—and have to implement the same solutions to each problem.

The richest man and poorest man on the planet wake up every day with the same exact problem: "I'm hungry."

There are good problems and there are bad problems: Good problems have solutions; bad problems do not have solutions.

Even in 2025, humans are living in the jungles of Ethiopia who walk around half-naked, eat tree branches and drink rain water, and pee and poop into a hole in the ground. They bathe in the river. The men have more than five wives who bear them many children. Each wife has her own mud hut. They are illiterate: They can't read or write. They do have a shared language and enjoy singing and dancing. They welcome tourists and entertain them by painting their faces with tribal insignia. They prefer to live like human animals in the jungle—than live like humanbeings in a civilization.

STARTREK: TO BOLDLY GO WHERE NO MAN HAS GONE BEFORE

Dr. Admiral Leonard H. McCoy
"Bones"
USS Enterprise

McCoy:
"We were speculating. Is God really out there?"

Captain James T. Kirk
USS Enterprise
23rd & 24th Centuries

Kirk:
"Maybe he's not out there, Bones.
Maybe he's right here: Human Heart."

LAWS OF INEXTRICABILITY
THE GRAND UNIFIED THEORY OF PHYSICS

1. PEACELESSNESS
2. HOMELESSNESS
3. LIFELESSNESS
4. HUMANBEINGLESSNESS
5. MORALITYLESSNESS
6. CHOICELESSNESS
7. LOVELESSNESS
8. MINDLESSNESS
9. HAPPYLESSNESS
10. MERCYLESSNESS
11. CREATIONLESSNESS
12. TRUTHLESSNESS
13. GODLESSNESS
14. RELIGIONLESSNESS
15. PROBLEMLESSNESS

16. LAWS OF INEXTRICABILITY
THE GRAND UNITED THEORY OF PHYSICS

PHYSICS:
ELECTRICITY and MAGNETISM are **INEXTRICABLY** interwoven into an interdependent matrix.

PARTICLES and WAVES are **INEXTRICABLY** interwoven into an interdependent matrix.

The FIELD THEORY OF PARTICLES and THEORY OF GRAVITY are **INEXTRICABLY** interwoven into an interdependent matrix.

QUANTUM FIELD THEORY and STRING THEORY: S-DUALITY are **INEXTRICABLY** interwoven into an interdependent matrix.

NEWTON F= MA
FORCE and ACCELERATION are **INEXTRICABLY** interwoven into an interdependent matrix.

EINSTEIN E =mc2
ENERGY and MATTER are **INEXTRICABLY** interwoven into an interdependent matrix.

ENGINEERING
INTERCHANGING VOLTAGE and CURRENT in an expression are **INEXTRICABLY** interwoven into an interdependent matrix.

INTERCHANGING FORCE (STRESS) and DEFORMATION (STRAIN) in an expression are **INEXTRICABLY** interwoven into an interdependent matrix.

LAWS OF INEXTRICABILITY
THE GRAND UNITED THEORY OF PHYSICS

BIOLOGY: METAMORPHOSIS
USE IT OR LOSE IT! are INEXTRICABLY interwoven into an interdependent matrix.

UMBILICAL CORD (USEFUL AND USELESS) is **INEXTRICABLY** interwoven into an interdependent matrix. **LAW OF INEXTRICABILTY:** CREATIONLESSNESS

NATURE and NURTURE are **INEXTRICABLY** interwoven into an interdependent matrix.

HUMAN ANIMAL and HUMANBEING are **INEXTRICABLY** interwoven into an interdependent matrix. **LAW OF INEXTRICABILTY:** HUMANBEINGLESSNESS

MATHEMATICS
CERTAINTY and UNCERTAINTY are **INEXTRICABLY** interwoven into an interdependent matrix.

RELIGION
PHYSICAL AND METAPHYSICAL WORLDS are **INEXTRICABLY** interwoven into an interdependent matrix. **LAW OF INEXTRICABILTY:** GODLESSNESS

GOOD and EVIL are **INEXTRICABLY** interwoven into an interdependent matrix. **LAW OF INEXTRICABILTY:** MORALITYLESSNESS (8 UNPREDICTABLE OUTCOMES)

BIOLOGY
The MIND and BODY are **INEXTRICABLY** interwoven into an interdependent matrix.

PHILOSOPHY
FREE WILL and DETERMINISM are **INEXTRICABLY** interwoven into an interdependent matrix.

PSYCHOLOGY
UNCONSCIOUS and CONSCIOUS are **INEXTRICABLY** interwoven into an interdependent matrix. **LAW OF INEXTRICABILTY:** MINDLESSNESS

LAWS OF INEXTRICABILITY
THE GRAND UNITED THEORY OF PHYSICS

LOVE and HATE are **INEXTRICABLY** interwoven into an interdependent matrix. **LAW OF INEXTRICABILTY:** LOVE**LESSNESS**

CARL JUNG
LIGHT and DARK are **INEXTRICABLY** interwoven into an interdependent matrix.

AUTONOMY and DEPENDENCY are **INEXTRICABLY** interwoven into an interdependent matrix.

SELF-INTEREST and ALTRUISM are **INEXTRICABLY** interwoven into an inter-dependent matrix.

COGNITIVE SCIENCE
OBJECTIVE VIBRATIONS and SUBJECTIVE PERCEPTION are **INEXTRICABLY** interwoven into an interdependent matrix.

HISTORY: CENTURIES OF REPEAT!
TODAY and YESTERDAY are **INEXTRICABLY** interwoven into an interdependent matrix.

EDUCATION
IGNORANCE and KNOWLEDGE are **INEXTRICABLY** interwoven into an interdependent matrix. **LAW OF INEXTRICABILTY:** MIND**LESSNESS**

WRITING
RIGHT BRAIN and LEFT BRAIN are **INEXTRICABLY** interwoven into an interde-pendent matrix. For instance: Right Brain: Creative & Left Brain: Editorial

CHINESE PHILOSOPHY OF **INEXTRICABILITY**
YIN and YANG are fundamental concepts in Chinese philosophy, particularly Taoism, representing complementary, interdependent, and opposing forces that interact to form a dynamic system and achieve balance and harmony.

The First Law
of Nature is
INEXTRICABILITY

Measure
the
Pleasure
Life is a Gift

Measure
the
Pain!
Life is a Punishment

15
Esoteric
Laws of
Inextricability

ONLY GOD
KNOWS
WHY?
AND GOD
KNOWS
BEST!

THE END OF
FEEL GOOD
PSYCHOBABBLE
TO NUMB
THE PAIN AND
KEEP THEM
SANE!

Q. Is Life a Gift or
a Punishment?

15
Absolute
Truths

Final Note

LAWS OF INEXTRICABILITY

If it was all up to me, I would hang a gargantuan, red, and electronic blinking billboard that reads:

WELCOME TO PLANET EARTH
WARNING: LAWS OF INEXTRICABILITY
CONDEMED! UNINHABITABLE!

Matter of fact, my acute sensitivity, is a personal suffering at times; and at other times a strength, if directed into my literary pursuits:

I, Sharon Esther Lampert, the **PHILOSOPHER QUEEN**, certainly, do not write to influence the reader in any way, shape, or form.

I write merely to communicate my perspicacious insights, and to share them with the readers.

If I also succeed in educating readers, then so be it, an added perk, as I cannot educate socialized human beings beyond their innate intelligence.

If I also succeed in enlightening readers, then so be it, another added bonus, as I cannot enlighten socialized human beings who prefer to live in a state of abysmal ignorance, deafening denial, and sheer stupidity.

If I also succeed in advancing and expanding the frontiers of human knowledge, then so be it, another added benefit of vast humanitarian dimensions.

FIND THE LIGHT AND LIVE IN THE LIGHT!

SHARON ESTHER LAMPERT
PHILOSOPHER QUEEN
V.E.S.S.E.L. : VERY. EXTRA. SPECIAL. SHARON. ESTHER. LAMPERT.
PRINCESS KADIMAH: 8TH PROPHETESS OF ISRAEL
PHOTON SUPERHERO OF EDUCATION: SMARTGRADES BRAIN POWER REVOLUTION

WHY? The Birth of a Philosopher Queen: Sharon Esther Lampert

At the age of nine, Sharon's **MOMMY** proclaimed, "My daughter is a poet, philosopher, and teacher. She is the Princess and the Pea! THE QUEEN HAS ARRIVED!"

As her vast imagination recalls, on the day she was born, her rough, tough, and tumble Holocaust survivor father was dumbfounded by her tears. How would his first blood relative, born after the Holocaust, a new big-blue-eyed bundle of brand new baby survive in the big-bad world? Moments after her arrival, he exclaimed, "She just came out of the womb, and she is already crying… I don't understand! Nothing has happened to her yet! WHY is she already crying? How is she going to survive? **Oy Veys Mir!**"

And later, she recalls telling him, "You of all people, a Holocaust survivor, the sole surviving member of your immediate family, already knew that this place called **Planet Earth** was a god-forsaken place of pain, suffering, and death: the violence is indigenous. WHY did you bring me here to the big bad world. WHY ME?"

Bittersweetly, it was very odd, how two very "shtarker" parents gave birth to a bouncing babe of extraordinary sensitivity and flaming intensity. How is it possible? WHY THEM?

Bittersweetly, so what if on the website, creativegenius.org, flaming intensity equals a creative genius. Anyhow, his little known fact does little to alleviate the inevitable mental anguish and suffering that is a natural consequence of this gifted reality of immortal genius.

Bittersweetly, during her childhood, she remembers answering WHY**?** to everything ever told to her:

Darkly, WHY do I have to kill other living beings, cook them, and eat them to live?
Darkly, WHY don't animals have to cook?
Darkly, WHY do I have to have a period for thirty-three years? WHY every single month? I am clearly in over my head!
Darkly, WHY isn't there a way out of this bloody mess, a pill or an operation?
Darkly, WHY is it too hot, to cold, too windy, too rainy to play outside?
Darkly, WHY do I have to catch a cold?
Darkly, WHY do I feel like I am going to die from the flu?
Darkly, WHY do I already need eyeglasses in the fourth grade?
Darkly, WHY do I have cavities?
Darkly, WHY do I need to wear dental braces for five years?

And I really don't want to wear cat-eyed glasses and dental braces and have wavy hair all at the same time!

Darkly, clear-plastic braces did not as yet exist.

Sweetly, wire rims, contact lenses, and hair dryers were in vogue.

Bittersweetly, she looked closely at her parents and wondered, "WHY do I know these people?"

Darkly, she looked at the snotty-nosed boys in her class, "WHY will I have to marry one of them?"

Darkly, in the fourth grade, her cat died and she cried for two whole years after the fact—and she still wants to know, "WHY did my cat have to die?

Darkly, WHY were all of my relatives murdered in the Holocaust? In the deepest depths of her soul, she thought, "First they force you to be born without your consent, and then to add insult to injury, they give you the business: racism, sexism, religious persecution, ageism.

Darkly, every five minutes you have to rise above a situation!

Darkly, WHY are you giving me the business?" was the quintessential WHY. WHY ME? is the concluding afterthought.

Darkly, WHY is all of this necessary?

What was considered normal was too hard! "NORMAL IS TOO HARD!" She exclaimed. A PHILOSOPHER QUEEN WAS BORN!

Enough already of her childhood!

At age 16, Sharon Esther left this note in the Wailing Wall in Jerusalem, Israel:

> Dear God:
> I am not asking for help.
> I am asking to be of help.
> I am at your service.
> Sharon Esther Lampert

The worst thing that has ever happened to me is that I was born.
The best thing that has ever happened to me is that I was born Jewish.

PHILOSOPHER QUEEN SHARON ESTHER LAMPERT

P.S. FIND THE LIGHT AND LIVE IN THE LIGHT!

Ancient Philosophers

* Aenisedemus
* Alcmaeon of Croton
* Alexander of Aphrodisias
* Alexander Polyhistor, Lucius Cornelius
* Anaxagoras
* Anaximander
* Anaximenes
* Antisthenes
* Arcesilaus
* Archelaus of Athens
* Archytas of Tarentum
* Aristotle
* Aurelius, Marcus
* Carneades
* Chrysippus
* Cicero, Marcus Tullius
* Cleanthes
* Confucius
* Critias
* Democritus
* Diogenes Laertius
* Diogenes of Apollonia
* Diogenes of Sinope
* Diotima of Mantinea
* Empedocles
* Empiricus, Sextus
* Epictetus
* Epicurus
* Euclides
* Heraclitus
* Hippias
* Hippon of Samos
* Lao Tzu
* Leucippus
* Lucretius
* Melissus of Samos
* Mencius
* Menippus
* Philo of Alexandria
* Philolaus
* Plato
* Protagoras
* Pyrrho
* Pythagoras
* Socrates
* Speusippus
* Sung-Tzu
* Thales
* Theophrastus
* Xenocrates
* Xenophanes
* Xenophon
* Zeno of Citium
* Zeno of Elea

Medieval Philosophers

* Abelard, Peter
* Albert of Saxony
* Alyngton, Robert
* Anselm of Canterbury
* Aquinas, Thomas
* Augustine of Hippo
* Auriol, Peter
* Averroes
* Avicenna
* Bacon, Roger
* Boethius, Anicius Manlius Severinus
* Buridan, John
* Damian, Peter
* Dante Alighieri
* Duns Scotus, John
* Eckhart, Meister Johannes
* Erfurt, Thomas of
* Eriugena, John Scotus
* Farabi
* Francis of Marchia
* Gersonides
* Ghazali
* Giles of Rome
* Godfrey of Fontaines
* Gregory of Rimini
* Herbert of Cherbury, Edward
* Heytesbury, William
* Holcot, Robert
* Hypatia
* Kilvington, Richard
* Kindi
* Magnus, Albertus
* Maimonides, Moses
* Marsilius of Inghen
* Nicholas of Autrecourt
* Ockham, William of
* Paul of Venice
* Penbygull, William
* Peter of Spain
* Plotinus
* Porphyry
* Richard the Sophister
* Roscelin, Johann
* Rufus of Cornwall, Richard
* Saadya Gaon
* Salisbury, John of
* Sharpe, Johannes
* William of Auvergne
* William of Champeaux
* William of Conches
* William of Moerbeke
* Wyclif, John

Renaissance Philosophers

* Agrippa, Heinrich Cornelius
* Argyropoulos, John
* Bacon, Francis
* Barbaro, Ermolao
* Bessarion, John
* Bodin, Jean
* Boehme, Jacob
* Bruni, Leonardo
* Bruno, Giordano
* Bude, Guillaume
* Burgersdijk, Franco
* Cajetan, Tommaso de Vio
* Campanella, Tommaso
* Cano, Melchior
* Cesalpino, Andrea
* Charron, Pierre
* Chrysoloras, Manuel
* Desgabets, Robert
* Eck, Johann
* Erasmus, Desiderius
* Ficino, Marsilio
* Galilei, Galileo
* Guicciardini, Francesco
* Landino, Cristoforo
* Las Casas, Bartoleme de
* Lipsius, Justus
* Luther, Martin
* Machiavelli, Niccolo
* Montaigne, Michel de
* More, Thomas
* Muret, Marc-Antoine de

* Nicholas of Cusa
* Oresme, Nicolas
* Palmieri, Matteo
* Pico della Mirandola, Giovanni
* Pomponazzi, Pietro
* Ramus, Peter
* Scaliger, Julius Caesar
* Suarez, Francisco
* Vair, Guillaume du
* Valla, Lorenzo
* Vanini, Giulio Cesare
* Vico, Giambattista
* Vitoria, Francisco de
* Vives, Juan Luis
* Zabarella, Jacopo

Early Modern Philosophers

* Arnauld, Antoine
* Astell, Mary
* Bayle, Pierre
* Beccaria, Cesare
* Bentham, Jeremy
* Berkeley, George
* Bolingbroke, Henry St. John
* Boyle, Robert
* Burke, Edmund
* Butler, Joseph
* Clarke, Samuel
* Collins, Anthony
* Condillac, Ettiene Bonnot de
* Cruz, Sor Juana Ines de la
* Cudworth, Ralph
* Descartes, Rene
* Diderot, Denis
* Elizabeth of Bohemia
* Gassendi, Pierre
* Geulincx, Arnold
* Hartley, David
* Herder, Johann Gottfried von
* Hobbes, Thomas
* Holbach, Paul-Henri Thiry, Baron d'
* Hume, David
* Jefferson, Thomas
* Johnson, Samuel
* Kant, Immanuel
* LeGrand, Antoine
* Leibniz, Gottfried Wilhelm
* Locke, John
* Malebranche, Nicolas
* Masham, Damaris Cudworth
* Mendelssohn, Moses
* Mersenne, Marin
* Montesquieu, Charles-Louis de Secondat, Baron de
* Newton, Isaac
* Paine, Thomas
* Pascal, Blaise
* Reid, Thomas
* Rousseau, Jean-Jacques
* Shaftesbury, Earl of
* Sidney, Algernon
* Smith, Adam
* Spinoza, Baruch
* Voltaire
* Wasa, Kristina
* Wollstonecraft, Mary

19th Century Philosophers

* Alcott, Amos
* Austin, John
* Bakunin, Mikhael
* Bastiat, Frederic
* Bauer, Bruno
* Bolzano, Bernard
* Boole, George
* Bosanquet, Bernard
* Bradley, Francis Herbert
* Brentano, Franz
* Caird, Edward
* Cantor, Georg
* Comte, August
* Constant, Benjamin
* Dilthey, Wilhelm
* Donoso Cortes, Juan
* Emerson, Ralph Waldo
* Engels, Friedrich
* Eucken, Rudolph
* Federov, Nikolai Fedorovich
* Feuerbach, Ludwig
* Fichte, Immanuel
* Fichte, Johann Gottlieb
* Frege, Gottlob
* Green, Thomas Hill
* Hamilton, William
* Hegel, G.W.F.
* James, William
* Kierkegaard, Soren
* MacColl, Hugh
* Marx, Karl
* Meinong, Alexius

* Mill, Harriet Taylor
* Mill, John Stuart
* Natorp, Paul
* Nietzsche, Friedrich
* Peirce, Benjamin
* Peirce, Charles Sanders
* Rosmini, Antonio
* Schelling, Friedrich
* Schleiermacher, Friedrich
* Schopenhauer, Arthur
* Spencer, Herbert
* Thoreau, Henry David
* Whewell, William

20th Century Philosophers

* Abramowski, Edward
* Adler, Alfred
* Adorno, Theodor
* Ajdukiewicz, Kazimierz
* Althusser, Louis
* Ambrose, Alice
* Anscombe, G.E.M.
* Arendt, Hannah
* Austin, J.L.
* Ayer, A.J.
* Bakhtin, Mikhail
* Barth, Karl
* Barthes, Roland
* Bataille, George
* Baudrillard, Jean
* Beauvoir, Simone de
* Berdyaev, Nicholai
* Bergmann, Gustav
* Berlin, Isaiah
* Blondel, Maurice
* Bohr, Niels
* Bonhoeffer, Dietrich
* Brouwer, L.E.J.
* Brzozowski, Stanislaw
* Buber, Martin
* Camus, Albert
* Carnap, Rudolf
* Cassirer, Ernst
* Chesterton, G.K.
* Chomsky, Noam
* Church, Alonzo
* Churchland, Paul
* Chwistek, Leon
* Croce, Benedetto
* Czezowski, Tadeusz
* Davidson, Donald
* Dawkins, Richard
* Deleuze, Gilles
* Dennett, Daniel
* Derrida, Jacques
* Dewey, John
* DuBois, W.E.B.
* Dummett, Michael
* Einstein, Albert
* Fanon, Frantz
* Feyerabend, Paul
* Fleck, Ludwig
* Foucault, Michel
* Gadamer, Hans-Georg
* Gettier, Edmund
* Gilligan, Carol
* Godel, Kurt
* Goodman, Nelson
* Habermas, Jurgen
* Hartshorne, Charles
* Hayek, Friedrich
* Heidegger, Martin
* Hempel, Carl Gustav
* Husserl, Edmund
* Ingarden, Roman
* Jaspers, Karl
* Kelles-Krauz, Kazimierz
* Kojève, Alexandre
* Kotarbinski, Tadeusz
* Kripke, Saul
* Kristeva, Julia

20th Century Philosophers

16 Famous Women Philosophers in History

* Krzywicki, Ludwik
* Kuhn, Thomas
* **Lampert, Sharon Esther**
* Langer, Suzanne
* Lesniewski, Stanislaw
* Levinas, Emmanuel
* Lewis, C.I.
* Lukasiewicz, Jan
* MacKinnon, Catharine
* Marcel, Gabriel
* Marcuse, Herbert
* Maritain, Jacques
* McTaggart, John
* Mead, George Herbert
* Merleau-Ponty, Maurice
* Moore, G.E.
* Natorp, Paul
* Nock, Albert Jay
* Nozick, Robert
* Oakeshott, Michael
* Ortega y Gasset, Jose
* Plantinga, Alvin
* Polanyi, Michael
* Popper, Karl
* Prior, Arthur
* Putnam, Hilary
* Quine, W.V.O.
* Rand, Ayn
* Rawls, John
* Reichenbach, Hans
* Ricoeur, Paul
* Rorty, Richard

* Russell, Bertrand
* Ryle, Gilbert
* Said, Edward
* Santayana, George
* Sartre, Jean-Paul
* Schlick, Moritz
* Schumpeter, Joseph
* Schutz, Alfred
* Searle, John
* Sellars, Wilfrid
* Shestov, Lev
* Strawson, Peter
* Tarski, Alfred
* Tatarkiewicz, Wladyslaw
* Tillich, Paul
* Turing, Alan
* Twardowski, Kazimierz
* Unamuno y Jugo, Miguel de
* Weber, Max
* Weil, Simone
* Whitehead, Alfred North
* Witkiewicz, Stanislaw
* Wittgenstein, Ludwig

1. Hypatia of Alexandria (c. 360–415 AD)
2. Émilie du Châtelet (1706–1749)
3. Mary Wollstonecraft (1759–1797)
4. Harriet Taylor Mill (1807–1858)
5. Simone de Beauvoir (1908–1986)
6. Elizabeth Anscombe (1919–2001)
7. Hannah Arendt (1906–1975)
8. Philippa Foot (1920–2010)
9. Ayn Rand (1905–1982)
10. Judith Butler (1956–)
11. Martha Nussbaum (1947–)
12. Iris Murdoch (1919–1999)
13. Mary Midgley (1919–2018)
14. Angela Davis (1944–)
15. Catherine MacKinnon (1946–)
16. Sharon Esther Lampert (1961-) PHILOSOPHER QUEEN

"Never Underestimate the Power of a Girl with a Book."

Ruth Bader Ginsburg

FAMOUS PHYSICISTS

Galileo Galilei

Galileo Galilei, an Italian mathematician, astronomer, and physicist, earned the title 'Father of Physics' due to his major contribution to finding the **Motion of Bodies** and the development of the telescope. Incidentally, the phases of Venus and also the four largest moons of Jupiter were discovered by him using his telescope. It is at the age of 19 that the pendulum's isochronal nature was discovered by him. Also, a swinging lamp's oscillations in the Pisa Cathedral were timed by him. It is Galileo's theories and concepts that formed the base for the laws of Motion and Theory of Relativity.

Isaac Newton

Sir Isaac Newton, associated with Cambridge University as a physicist and mathematician, became famous after propounding **Three Laws of Motion** that established a connection between objects and motion. The **Law of Universal Gravitation** was discovered by him and it is this law that defined the behavior of falling bodies. Also, he invented calculus in mathematics.

Albert Einstein

With the **Theory of Relativity**, Albert Einstein transformed the landscape of physics. Also, this theory majorly contributed to designing the theory of Quantum Mechanics. The mass-energy equivalence formula (**E=mc2**), propounded by Albert Einstein etched his name as the greatest physicist of all time in the books of history. Because of his contributions to modern science, the Nobel Prize was awarded to him in 1921.

Sharon Esther Lampert - **Published 80+ Books - Answered 50+ Questions**

Physicist: **LAWS OF INEXTRICABILITY: THE GRAND UNIFIED THEORY OF PHYSICS**
Psychobiologist: SPERM MANIFESTO: THERE IS ONLY ROOM FOR **ONE** AT THE TOP
Prodigy: 10 ESOTERIC LAWS OF GENIUS AND CREATIVITY
Paladin of Education: SMARTGRADES BRAIN POWER REVOLUTION
Poet: **POETRY WORLD RECORD**
Poet: AWESOME ART OF ALLITERATION USING ONE LETTER OF THE ALPHABET
Poet: The Greatest Poems Ever Written on Extraordinary World Events
Prophet: **GOD IS GO! DO!** THE 22 COMMANDMENTS

What Happens When You Dress Up Albert Einstein as Marilyn Monroe?

SHARON ESTHER LAMPERT

PHENOMENON
SCIENTIST, ARTIST, EDUCATOR, THEOLOGIAN
Published: 80 BOOKS - Answered: 50 Questions
Filmmaker: 50+ "One Nominated Film"

PRODIGY
UNLEASH THE CREATOR
THE GOD WITHIN
10 ESOTERIC LAWS OF
GENIUS AND CREATIVITY

PROPHET
WHO KNEW GOD WAS
SUCH A CHATTERBOX
GOD IS GO! DO!
THE 22 COMMANDMENTS
A UNIVERSAL MORAL COMPASS

PHYSICIST
NOW YOU KNOW
EVERYTHING ABOUT
EVERYTHING: LAWS
OF INEXTRICABILITY

PSYCHOBIOLOGIST
SPERM MANIFESTO
THERE IS ONLY ROOM
FOR ONE AT THE TOP

PHILOSOPHER QUEEN
WOMEN HAVE
ALL THE POWER

THE PHILOSOPHY
OF LOVE

THE PHILOSOPHY
OF EVIL: THE
DOUBLE WHAMMY

PRODUCER
80+ Short Films

PSYCHIATRIST
14 Relationship Stratigies

GENIUS: EXODUS 31: 1-3
THE GIFT OF DIVINE REVELATION

www.SharonEstherLampert.com
FANS@SharonEstherLampert.com

POET (25+ Books)
WORLD POETRY RECORD

THE GREATEST POEMS
EVER WRITTEN ON
EXTRAORDINARY
WORLD EVENTS

THE AWESOME ART OF
ALLITERATION USING
ONE LETTER OF THE
ALPHABET (6 Books)

FIRST WOMAN TO WRITE
BOOK ON JEWISH HISTORY
IN 5 MINUTES—4 REFRAINS

#1 POETRY WEBSITE FOR
STUDENT PROJECTS

**PALADIN OF EDUCATION
(25+ Books)**

SMARTGRADES
BRAIN POWER
REVOLUTION

40 UNIVERSAL
GOLD STANDARDS
OF EDUCATION

HOW DOES
LEARNING
TAKE PLACE?

8 GOALPOSTS OF
EDUCATION

PINUP
SEXIEST CREATIVE
GENIUS IN
HUMAN HISTORY

What Happens When You Dress Up Albert Einstein as Marilyn Monroe?

SHARON ESTHER LAMPERT

Prodigy
Prophet
Philosopher
Physicist
Psychobiologist—Rockefeller University
Peacemaker
Poet
Paladin of Education
PHOTON SUPERHERO OF EDUCATION
Princess Kadimah: 8TH PROPHETESS OF ISRAEL
Piano-**P**laying Cat
President
Publisher
Playdate
Painter
Photographer
Princess & **P**ea
Performer:Vocalist
Player: Jock
Phoenix
PINUP

Honored Sharon Lampert
with an Award for
"Multi-Interdisciplinary Studies"
YOUTUBE video

WEBSITES
SharonEstherLampert.com
WorldFamousPoems.com: #1 Poetry Website for Students
Schmaltzy.com: Piano-Playing Cat
TrueLoveBurnsEternal.com

PUBLISH
WritersRunTheWorld.com
PalmBeachBookPublisher.com

EDUCATION
Smartgrades.com
BooksNotBombs.com

YOU HAD TO OUTDO MOSES

THE 22 COMMANDMENTS

All You Will Ever Need to Know About God

A Universal Moral Compass for All People
For All People, For All Religions, For All Time

1. **LIFE** Over Death

2. **STRENGTH** Over Weakness

3. **DEED** Over Sin

4. **LOVE** Over Hatred

5. **TRUTH** Over Lie

6. **COURAGE** Over Fear

7. **OPTIMISM** Over Pessimism

8. **SHARING** Over Selfishness

9. **PRAISE** Over Criticism

10. **LOYALTY** Over Abandonment

11. **RESPONSIBILITY** Over Blame

12. **GRATITUDE** Over Grievances

13. **REWARD** Over Punishment

14. **ALLIES** Over Enemies

15. **CREATION** Over Destruction

16. **EDUCATION** Over Ignorance

17. **COOPERATION** Over Competition

18. **FREEDOM** Over Oppression

19. **COMPASSION** Over Indifference

20. **FORGIVENESS** Over Revenge

21. **PEACE** Over War

22. **JOY** Over Suffering

By Sharon Esther Lampert
Princess Kadimah, 8TH Prophetess of Israel
Read My Book: **"THE 22 COMMANDMENTS"**

GENIUS: GIFTS OF DIVINE REVELATION

Published 80+ Books

Prodigy—Princess & Pea!
UNLEASH THE CREATOR THE GOD WITHIN: 10 ESOTERIC LAWS OF GENIUS AND CREATIVITY
AWESOME ART OF ALLITERATION USING ONE LETTER OF THE ALPHABET (6 Books)

Prophet
WHO KNEW GOD WAS SUCH A CHATTERBOX: A WORKING DEFINITION OF GOD—GOD IS GO! DO!
22 COMMANDMENTS: ALL YOU WILL EVER NEED TO KNOW ABOUT GOD

Physicist
NOW YOU KNOW EVERYTHING ABOUT EVERYTHING: LAWS OF INEXTRICABILITY

Psychobiologist
SPERM MANIFESTO: THERE IS ONLY ROOM FOR ONE AT THE TOP

Philosopher Queen
WOMEN HAVE ALL THE POWER BUT HAVE NEVER LEARNED HOW TO USE IT
THE PHILOSOPHY OF LOVE: SELF-LOVE & BONUS LOVE
THE PHILOSOPHY OF EVIL: THE DOUBLE WHAMMY

Paladin of Education (+25 Books)
PHOTON SUPERHERO OF EDUCATION
SMARTGRADES BRAIN POWER REVOLUTION
- THE SILENT CRISIS DESTROYING AMERICAS BRIGHTEST MINDS—BOOK OF THE MONTH
- EVERY DAY AN EASY A!
- 40 UNIVERSAL GOLD STANDARDS OF EDUCATION
- HOW DOES LEARNING TAKE PLACE

Poet (+25 Books)
ONE OF THE WORLD'S GREATEST POETS
POETRY WORLD RECORD: 120 WORDS OF RHYME
THE GREATEST POEMS EVER WRITTEN ON EXTRAORDINARY WORLD EVENTS
HTTP://FAMOUSPOETSANDPOEMS.COM/POETS.HTML

Princess Kadimah
8TH PROPHETESS OF ISRAEL: THE 22 COMMANDMENTS

PINUP
SEXIEST CREATIVE GENIUS IN HUMAN HISTORY

Artists March to the Beat of a Different Drummer
Sharon Esther Lampert Marches to the Beat of an Entire Orchestra

Poet, Philosopher, Prophet, Peacemaker
Paladin of Education, Princess & Pea
Phoenix, PHOTON, PINUP, Prodigy

Blue-Eyed. Brilliant. Beautiful. Buxom. Books. Blessed.

Sharon Esther Lampert was born an **OLD SOUL**—She was never young! Sharon is a lefty.

At age nine, her mother declared: "My daughter is a poet, philosopher, and teacher!" She nicknamed her daughter, "The Princess and Pea!" **"THE QUEEN HAS ARRIVED!"**

Sharon's greatest literary works woke her up in the middle of the night—and made her get up out of bed—and write them down. Sharon writes an entire book in one day!

Sharon's mother was the sole person in Sharon's life who knew who she was from the **INSIDE OUT**! Her beloved mother also knew to her very last breath... the exact day and to the minute when she would die! (Eve Paikoff Lampert: June 3, 1925—May 5, 1985).

Later in life, Sharon purchases a **green-pea pendant**, at the Broadway show, **"Once Upon a Mattress"** starring Sarah Jessica Parker. She wore the **green pea** every day around her neck with a beautiful Jewish-star pendant from Haifa, Israel.

Sharon Esther's Gifts Are Metaphysical—Beyond the Scope of Scientific Inquiry

There Are No Rough Drafts!—Revelation: The Books Write Themselves!

Sharon Has 4 Books with GOD in the Title

"A LIST" Sharon Esther Lampert Is One of the World's Greatest Poets
http://famouspoetsandpoems.com/poets.html

#1 Poetry Website for Student Poetry Projects

On a global scale, Sharon's poetry is used by teachers for their poetry lesson plans, and by students for their poetry school projects (see **student FAN MAIL**).

New York University Awards — YouTube Videos

Sharon Esther earned three degrees from **NYU**—and she was honored with two NYU awards. Sharon represented her class at her graduation—and was honored with an award for "Multi-Interdisciplinary Studies." She also played on the **NYU** Women's Varsity Basketball team as a center in the $16-million Coles Sports Center. Sharon won an "**NYU** Weightlifting Contest—Sharon was the sole contestant—so she won! (**Washington Square News** article).

PINUP

SEXIEST CREATIVE GENIUS IN HUMAN HISTORY

Sharon Esther Lampert

MYLIFE Is an OPENBOOK to KNOWME Is to README

1.
PRODIGY
10 Esoteric Laws of
Genius & Creativity

Awesome **A**rt of
Alliteration Using
One Letter of The
Alphabet

2.
PROPHET
22 COMMANDMENTS
A Universal Moral
Compass For All People

Who Knew **GOD** Was
Such a Chatterbox
A WORKING DEFINITION OF GO
GOD IS GO! DO!

3.
POET
The Greatest Poems
Ever Written on
Extraordinary
World Events

POETRY WORLD RECORD

4.
PHYSICIST
LAWS OF INEXTRICABILIT

5.
PSYCHOBIOLOGIST
SPERM MANIFESTO

6.
PHILOSOPHER
WOMEN HAVE ALL THE POW
PHILOSOPHY OF LOVE
PHILOSOPHY OF EVIL

7.
**PHOTON SUPERHERO &
PALADIN OF EDUCATION**
**SMARTGRADES
BRAIN POWER REVOLUTIO**

A PHENOMENON...
SHARON ESTHER LAMPERT

Lithe and lovely ... like a fawn.
This lady fascinates me ... from dusk till dawn.
Feminine and comely ... she's beyond belief
A blue-beam from her eyes ... is my soothing relief.

Girlish in her braces ... maidenly in her style
I yearn for her embraces ... and adore her friendly smile.
As tasteful as any artist ... you'll ever see
She's a compendium of class ... from A to Z.

If you'd like to see a figure, that puts Venus to shame
Behold her in a swimsuit, and your passions will aflame.
Ever exuding goodness . . . guided from above
Miss Sharon is the essence, and epitome of Love.

She's the inspiration of sages, and also fools like me
And the most magnificent female, I'm sure I'll ever see.
The nights are now endearing, & never filled with doubt
I sometimes wake up singing, cause it's Sharon . . .
I dream about.

Affectionately, . .
A devoted fan,
—Harry McVeety

Dear Sharon,

You are not only an exquisite poet, you're beautiful! Am smitten by your luminous beingness. Are you an angel in disguise--a so-called malachim in Hebrew if I am not mistaken.

Thank you for your wondeful open-hearted response.

Your photo will sit next to those of Gautama Buddha and the Blessed Virgin Mary.

I will follow your sound esoteric advise regarding the positioning of your photo and the two other icons. I am deeply impressed that you are very conscious about the concept of sacred space and the flow of spiritual energy. So please send me your precious photo as soon as possible.

P.S. Will you be generous enough to send me your signed photo which I will place on the secret altar of my heart, lit by the menorah, the seven-stemmed candelabra of your inspiration, O mystical muse, O Rose of Sharon...

Your ardent fan and admirer,
— Felix Fojas, the cybercat with a mystical meow
Chico, CA, 95926

KADIMAH PRESS: *Gifts of Genius*

Divine Relevations: My Books Write Themselves!

25+ POETRY BOOKS
Poet: The Greatest Poems Ever Written on Extraordinary World Events
Title: I Stole All the Words from the Dictionary
ISBN Hardcover: 978-1-885872-06-7
ISBN Paperback: 978-1-885872-07-4
ISBN E-Book: 978-1-885872-08-1

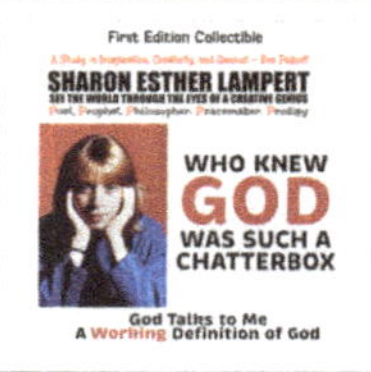

Prophet
Title: GOD TALKS TO ME: A WORKING DEFINITION OF GOD GOD IS GO! DO!
WHO KNEW GOD WAS SUCH A CHATTERBOX?
ISBN Hardcover: 978-1-885872-33-3
ISBN Paperback: 978-1-885872-34-0
ISBN E-Book: 978-1-885872-36-4

Prophet
Title: The 22 Commandments: All You Will Ever Need to Know About God
A Universal Moral Compass For All People, For All Religions, For All Time
ISBN Hardcover: 978-1-885872-03-6
ISBN Paperback: 978-1-885872-04-3
ISBN E-Book: 978-1-885872-05-0

Psychobiologist
Title: SPERM MANIFESTO: THERE IS ONLY ROOM FOR ONE AT THE TOP
ISBN Hardcover: 979-8-3305-2236-1
ISBN Paperback: 979-8-3305-2238-5
ISBN e-book: 979-8-3305-2237-8

Prodigy
Title: Unleash the Creator The God Within
10 Esoteric Laws of Genius and Creativity
ISBN Hardcover: 978-1-885872-21-0
ISBN Paperback: 978-1-885872-22-7
ISBN E-Book: 978-1-885872-23-4

Published 80+ Books

Prodigy
AWESOME ART OF ALLITERATION USING ONE LETTER OF THE ALPHABET
Title: CUPID: Language of Love—Written in Letter C
ISBN Hardcover: 978-1-885872-55-5
ISBN Paperback: 978-1-885872-56-2
ISBN E-Book: 978-1-885872-57-9
(6 Books in Collection: Letter C, D, S, P1, P2 and T)

Popular: Children's Book, Ages 8-12
Title: SCHMALTZY: IN AMERICA, EVEN A CAT CAN HAVE A DREAM
TRUE STORY! WORLD FAMOUS PIANO-PLAYING CAT
ISBN Hardcover: 978-1-885872-39-5
ISBN Paperback: 978-1-885872-38-8
ISBN E-Book: 978-1-885872-37-1
Schmaltzy.com

First Book of
COLOR-CODED
Vocabulary Words

Popular
Title: SILLY LITTLE BOYS: 40 RULES OF MANHOOD
Q: How Do Silly Little Boys Grow into Sane Big Men?
14 Global Catastrophes of Violence Against Women
ISBN Hardcover: 978-1-885872-29-6
ISBN Paperback: 978-1-885872-35-7
ISBN E-Book: 978-1-885872-41-8

Popular: Every Great Relationship Begins with the Perfect Meal
Title: SEX ON A PLATE: FOOD AS FOREPLAY
THE COOKBOOK OF EVERLASTING LOVE
ISBN Hardcover: 978-1-885872-46-3
ISBN Paperback: 978-1-885872-48-7
ISBN E-Book: 978-1-885872-47-0
TrueLoveBurnsEternal.com

Blessing 1. My Genetics, Two Sets of Artistic Genes and Gift of Genius: LEFTY—Like MOMMY!
• Genetic Inheritance: Painter Maternal Grandfather Benjamin Paikoff &
  Sculptor Father Abraham Lampert (Exhibit in Museum of Jewish Heritage, NYC)
• Vocalist: Ashira Orchestra, 18 Years: Ramaz Women's Service (YouTube video)
• Athlete: "Faster Than Any Boy, Anytime, Anywhere, Any Age!"

Blessing 2. My Life: Dawn of Digital Revolution
• APPLE: The Golden Age of Personal Computers
• ADOBE: The Golden Age of Creativity
• INGRAM: The Golden Age of Publishing
• SOCIAL MEDIA: The Golden Age of Internet & Global Communication
• iTUNES: The Golden Age of Music and Lyrics

Blessing 3. My Loved Ones
• **SELF-LOVE:** "Mindfulness, Meditation, Mantra, and Music Mitigates Madness!"
• Unconditional True Love: My MOMMY Eve Paikoff Lampert
• My PURRfect Children: SCHMALTZY and FALAFEL, Schmaltzy.com
• My "Friends First and Forever, and Family" NYU Tisch Professor Karl Bardosh
• My Metaphysical Sister: Poet on a Mission Hannah Sezenes, "ELI, ELI"
• My 7 Practice Husbands, Artist—Muses, Dates, and NYC Night Life
• My Bubbe Esther Tulkoff, EstherTulkoff.com

Blessing 4. My NYU Education, Educators, and Awards
My Solomon Schecter Day School, Jewish Theological Seminary of America
My NYU Education: B.A., M.A., M.A. and AWARDS (YouTube video)
• NYU Professor Laurin Raiken
  NYU "Multi-Interdisciplinary Award" and M.A. Class Representative at Graduation
• Rockefeller University, NYC, Publication: "Hyperphagia and Obesity Induced by
  Neuropeptide Y" — Lab of Dr. Sarah Leibowitz and Dr. Glen Stanley
• AWARD: 100-Year Scholarship Award Winner, Presented by NYC Mayor Edward Koch
• AWARD: Empire Science Scholarship Award Winner
• AWARD: Jerusalem Fellowship Award, Aish Hatorah, Israel
• AWARD: First Prize: Upper East Side Resident Writing Contest
• Voice Teachers: Andy Anselmo of The Singer's Forum, NYC & Estelle Leibling
• Cantor Sherwin Goffin of LSS & Riva Alper of RAMAZ Women's Service (18 Years)

Blessing 5. My Sports
• NYC Marathon
• Basketball: NYU Women's Varsity Basketball Team, Center, NYU Coach Sherri Pickard
• NYU Weightlifting Contest Winner! $16 Million Coles Sports Center
  (solo contestant—so I won!)
• Basketball: NYC Urban Professional League
• B-Ball Coaches Chicago Bulls Phil Jackson and Boston Celtics Bill Walton,
  Omega Institute, NY
• Basketball and Softball—Coach Sandy Pyonin
• Skiing: Heavenly, Lake Tahoe, Nevada
• Tennis: NYC Central Park Tennis Courts
• Baseball, Hall of Fame: Coaches Wilma Briggs and Jean Harding, Omega Institute, NY

Blessing 6. My Inspirations
• ISRAEL: "**AM YISRAEL CHAI!**"
  Lambs to Slaughter to Lions & Light of the World: 22% of Nobel Prizes!
• NYC: The Golden Age of Personal Freedom & Creative Self-Expression
• AMERICA: Land of Unlimited Possibility, Potential, and Prosperity!

NYU Gallatin Professor
Laurin Raiken and Me at
My NYU M.A. Graduation

NYU Tisch Professor
Karl Bardosh and Me
Friends First &
Family Forever!

LITERATURE IS POWERFUL BEYOND WORDS FOR IT CREATES WORLDS

Sharon Esther Lampert

EVERY THOUGHT IN YOUR HEAD WAS PUT THERE BY A WRITER

Sharon Esther Lampert

Please Keep in Touch!

Website, Facebook, Twitter, Instagram, YOUTUBE, Pinterest

FANS@SharonEstherLampert.com

Published 80+ Books

GENIUS: GIFTS OF DIVINE REVELATION

MY BOOKS WRITE THEMSELVES!

I Am Mortal
MY BOOKS ARE IMMORTAL
Please Handle My Books Gently
My Books Are My Remains

This book was compiled in four parts:
Part 1. Divine Revelation: February 9th, 1999
Part 2. Essays: September 22, 2009
Part 3. Publish: 2023 GOD OF WHAT?
Part 4. Publish: 2025 NOW YOU KNOW EVERYTHING

SHARON ESTHER LAMPERT
SEE THE WORLD THROUGH THE EYES OF A CREATIVE GENIUS

FANS@SharonEstherLampert.com